I0049956

APICULTURE
FOR THE
21ST CENTURY

**ROGER HOOPINGARNER
AND
LAWRENCE CONNOR**

EDITORS

Wicwas Press
Cheshire Connecticut USA

First Edition

Published by Wicwas Press, LLC, P.O. Box 817, Cheshire, CT 06410-0817 USA

This book, or parts thereof, must not be reproduced in any form without permission in writing from the publisher except by a reviewer to quote brief passages in a review of the book.

Library of Congress Cataloging-in-Publications Data

ISBN: 1-878075-10-1, Limited Edition with linen cover for Symposium Participants and Collectors
ISBN: 1-878075-11-X, General Edition

Preface

ROGER HOOPINGARNER

Like most people, when I decided to retire from Michigan State University I had many mixed emotions. Very reluctant to leave the Department that I had been associated with for 38 years, and I had found many friends within and outside of the beekeeping circle. So leaving these things that I enjoyed was tough. One of the things that I did not want was a retirement dinner where everyone either says a lot of nice things or they give a "roast." However, the Chairperson at the time, Dr. J. Mark Scriber, suggested what I thought would be a great leaving gift. The suggestion was to have a symposium that I would design.

I have been to many symposia over the years and had developed many ideas regarding the best format. Unfortunately, I found out that time and other constraints make the "perfect" symposium probably just a dream. I decided that a two-day format with about four topics was about all that most people would like and this provided my first constraint. What would be those four topics? I had previously decided on the title of "Apiculture for the 21st Century" so that did guide me somewhat. I tried to think of what were the most pressing subjects, as well as those that would be the most topical for the first years of the next century. The limit of numbers caused me to drop such topics as pollination even though this was a subject that was very probably closest to my main research interest—though also not a subject that has pressing research needs. When I had selected the topics, then I picked four section leaders to then select the four, or five, leaders in their fields to present research discussions for the symposium.

The selection part was fun. What I learned after that time suggested it would have been far easier to slip off into retirement as quietly as possible. I certainly now want to express my sincere thanks to Linda Gallagher of the Entomology Department and the rest of the staff for all their help in getting out the final result. The symposium was obviously a modern one since it never would have happened without that wonderful communicator e-mail. Getting answers from participants in just a matter of hours certainly aided in getting the symposium presented in a relatively short time.

I want to again thank all the participants who took time right out of the middle of their research seasons to send me off into retirement in such grand fashion. Since I decided to retire at the end of June 1996 the timing was not the best for active researchers to come to East Lansing, Michigan. However, not only did they come but also they even stayed long enough so that the closing banquet was even the mild "roast" that I decided that I did not want. They were all wonderful and I had the night of my life. I cannot leave the research-teaching world of apiculture without a little comment on what I think is its future—the real reason for this volume. I sometimes wish I had been born a little later as computers have opened up a whole new era for discovery, cooperation and the development of knowledge. Sitting at this desk I can access almost all of the world's literature on honeybees and keep up to date on research everywhere—marvelous tool for any scholar. With that "on-line" ability I hope to stay current well into the 21st century. My present bibliography has close to 10,000 references and most of them with abstracts. I could not have even dreamed of that knowledge base when I started research 40 plus years ago.

Finally, I want to thank those marvelous little ladies, the honey bees, for giving me so much joy for these past 50 years.   •

List of participants

Ahmad Alghandi
Michigan State University

Dr. Robert Arther
Bayer Animal Health

Dr. George Ayers
Michigan State University

David/Blanche Barber
Davisburg, MI

Puja Batra
Michigan State University

Versal Beemer
Freesail, MI

Andrew Banczyk
Manton, MI

Richard Buck
Richard@acctours.com)

Ian Burns
Roseville, TN

Dewey Caron
Newark, DE

Sue Cobey
Ohio State University

Larry Connor
Cheshire, CT

Miguel Drechavaleta
MEXICO

Sydney D'Silva
Michigan State University

Dave Dagostino
Highland Township, MI

Colleene Davidson
Marion, MI

Elaine Evans
St. Paul MN

Donald Eveleth
Adrian, MI

Jennifer Fewell
Arizona State University

Kim Flottum
Medina, OH

Mary Folkema
Fremonte, MI

Vehaley Gathenko
West Bend, WI

Tom or Suki Glenn
Fallbrook, CA

Ed Grafius
Michigan State
University

Dan & Joyce Guthrie
Utica, MI

Soes Hadisoesiio
University of Guelph

Harlan Hall
Gainesville, FL

Dr. H. Hamel
Bayer Animal Health

Murray Hanna
Okemos, MI

Mike Hansen
Michigan Dept. of
Agriculture

John Harbo
Baton Rouge, LA

Larry&Ida Hasselman
Fremont MI 49412

Matt Higdon
Hallsville, MO

John & Mildred Hogg
Galesburg, MI

Zachary Huang
Champaign, IL

Gregg Hunt
Purdue University

Paul Johnson
Benton Harbor, MI

Deborah Kalicin
Lisbon, NY

Ron Katsma
Plainwell, MI

Terry Klein
St. Charles, MI

Dr. Nikolaus Koeniger
Oberusel, GERMANY

Jasna Kralj
Guelph, Ontario

Sharon & Ray Kusmal
Clinton, MI

Terry Lee Lansing
Ann Arbor, MI

Myron Lindahl
Harmantown, MN

Larry Lindokken
Mt. Horeb, WI

Gerald M. Loper
Tucson AZ

Bill Loyd & Elizabeth Wagner
Ann Arbor, MI

Kay Marchioni
Cincinnati, OH

Daniel Joseph Markey
Belleville, MI

Steve McDaniel
Manchester MD

Doug McRory
Guelph Ontario

Norberto Milani
Udine, ITALY

Jim & Naomi Miller
Michigan State University

Medhat Nasr
Guelph Ontario

Ronald Ochoa
Ann Arbor, MI

Dr. Gard Otis
Guelph, Ontario

Robert Page
Davis, CA

Warren Parsons
Jackson, MI

Anagela Pishny
Bayer Animal Health

Deanna Prough
Utica, MI

Homer Pugh
Troy, MI

Ken Richardson
Brooklyn, MI

Lloyd & Judy Schmaltz
Clarkston, MI

Joerg Schmidt-Baily
Frankfurt, GERMANY

Dr. Mark Scriber
Michigan State University

Ken Sharlow
Homer, MI

Karen Shaw
Livonia, MI

William & Barbara Sirr
Berkley, MI

Deborah Smith
Lawrence, KS

Maria Spivak
St. Paul, MN

Fred & Mary Ann Stehr
Michigan State University

Rhonda Struble
Bloomfield Hills, MI

Roger Sutherland
Ann Arbor, MI

John & Mary Lou Tanton
Petoskey, MI

O.R. Taylor
Lawrence, KS

Richard Taylor
Interlaken NY

Eva Toth
Mishawaka, IN

William Towne
Kutztown, PA

Mr. & Mrs. Jack Turner
Farmington Hills, MI

Gary Veale
Freeport, MI

Kirk Visscher
Riverside, CA

Keryn Wilkes
Coogee, AUSTRALIA

John P. Wrosch
Ann Arbor, MI

Presenters

Dr. O. R. Taylor, Jr.
Department of Entomology
7005 Haworth
University of Kansas
Lawrence, KS 66045-2106
monarch@falcon.cc.ukans.edu

Dr. Gerald Loper
Carl Hayden Bee Research
Center
2000 E. Allen Rd.
Tucson, AZ 85719
gmlop@tucson.ars.ag.gov

Dr. Glenn Hall
Department of Entomology
Building 970
University of Florida
Gainesville, FL 32611-0740
hgh@gnv.ifas.uflf.edu

Dr. Gard Otis
Department of Environ. Biology
University of Guelph
Guelph, Ontario N1G 2W1
CANADA
GOTIS@evbhort.uoguelph.ca

Dr. Nikolaus Koeniger
Institut fuer Bienenkunde
Karl von Frisch-Weg 2
D-61440 Oberursel
GERMANY
Nikolaus.Koeniger@em.uni-
frankfurt.de

Dr. Norberto Milani
Dipartimento di Biologia
University degli Studi di Udine
Via delle Scienze, 208
I 33100 Udine, ITALY

Dr. Diana Sammataro
Dept. of Entomology
Penn. State University
University Park PA

Jorg Schmidt-Bailey
Department of Entomology
Rutgers, The State University of NJ
93 Lipman Drive, Blake Hall
New Brunswick NJ 08901-8524
joerg@aesop.rutgers.edu

Dr. Marla Spivak
Department of Entomology
219 Hodson Hall, 1980 Fowell
University of Minnesota
St. Paul, MN 55108
spiva001@maroon.tc.umn.edu

Dr. Robert Page, Jr. *
Department of Entomology
University of California
Davis, CA 95616-8584
repage@ucdavis.edu

Dr. Jennifer Fewell
Department of Zoology
Arizona State University
Tempe, AZ 85287-1501
j.fewell@asu.edu

Dr. John Harbo
USDA Bee Breeding Laboratory
1157 Ben Hur Rd.
Baton Rouge, LA 70820
jharbo@asrr.arsusda.gov

Dr. Greg Hunt
Department of Entomology
Purdue University
West Layfayette, IN 47907-1158

Dr. Brian Smith *
Department of Entomology
1735 Neil Ave.
Ohio State University
Columbus, OH 43210-1220
bsmith@magnus.acs.ohio-state.edu

Dr. Ernesto Guzmán-Novoa
Santa Cruz 29-B
Las Haciendas
52140 Metepec
Edo. de Mexico MEXICO

Dr. Fred Dyer *
Department of Zoology
203 Natural Science
Michigan State University
East Lansing, MI 48824
fcdyer@pilot.msu.edu

Dr. Deborah R. Smith
Department of Entomology
7005 Haworth
University of Kansas
Lawrence, KS 66045-2106
dsmith@ukanvax

Dr. Kirk Visscher
Department of Entomology
University of California
Riverside, CA 92521
visscher@citrus.ucr.edu

Dr. Z-Y Huang
Department of Entomology
Michigan State University
East Lansing MI 48824
Phone 517-353-8136
Fax 517-353-4354
bees@msu.edu

Roger Hoopingarner
Department of Entomology
Michigan State University
East Lansing, MI 48824-1115
Phone 517-353-8136
Fax 517-353-4354
roghoopy@pilot.msu.edu

* Presentation not included in this volume.

Apiculture For the 21st Century

Table of Contents

The value of single-drone inseminations in selective breeding of honey bees

JOHN R. HARBO
HONEY BEE BREEDING, GENETICS, & PHYSIOLOGY LABORATORY
AGRICULTURAL RESEARCH SERVICE, USDA
1157 BEN HUR ROAD, BATON ROUGE, LA 70820

phone: 225-767-9288
fax: 225-766-9212
email: jharbo@asrr.arsusda.gov

Abstract

By mating queens with spermatozoa from a single drone, it was possible to increase the expression of a colony trait that was found only at low levels in colonies with queens mated to many drones. Nonreproduction of varroa mites averaged 22% in colonies with queens mated to one drone but only 6% in colonies with queens mated to six drones. Mating with a single drone increased the phenotypic variance among a group of colonies, thus making it possible to detect a colony characteristic that was masked by multiple mating. The survival and fecundity of singly mated queens that were less than four months old were not different from those of sister queens that were each mated to six drones. However, singly mated queens produced fewer progeny in a test that began with queens that were 7 months old. Therefore, the technique of using singly mated queens may be useful in field testing and selective breeding as long as the queens are less than about six months old during the evaluation.

Introduction

Plans for the selective breeding of honey bees are slightly different from those of most other plants and animals because all spermatozoa produced by a male honey bee are genetically identical. A normal male honey bee (drone) is haploid, and in the production of spermatozoa a drone produces about 10 million replicates of the gamete (an unfertilized egg) from which he developed.

The technique of inseminating a queen with a single drone is not new. It was made possible by the development of instrumental insemination of queen bees and was described by Mackensen & Roberts (1948). When a queen is mated to one drone, worker bees in the colony all have identical genetic material from their father, who is represented by identical spermatozoa that are now in the sperm storage organ (spermatheca) of the queen. Thus the worker bees are more closely related than normal sisters and have a relatedness of 0.75. A group of closely related sisters in a colony (the daughters of the same drone) is called a subfamily.

Walter Rothenbuhler made extensive use of the single-drone insemination and was probably its strongest advocate. "If a colony is composed of only one subfamily, it expresses the full intensity of each genetically determined behavior characteristic... unmodified by social environmental factors imposed by bees of other subfamilies" (Rothenbuhler 1960).

In nature a queen mates with many drones. A queen probably retains some of the spermatozoa from each of her 10-20 matings, and the resulting colony of bees consists of many subfamilies, one from each of the drones that mated with the queen. Therefore, when a queen is mated to many drones, the worker bees in that colony have much more genetic diversity than do the workers in a colony with a queen inseminated with a single-drone. In most cases this diversity is probably beneficial for the colony (Oldroyd *et al.* 1992, Fuchs and Schade 1994). Page *et al.* (1995) conclude that colonies with greater genetic diversity are more average with respect to most characteristics (such as bee population, cell size, defense behavior, diseases, mortality rate, etc.) and that being average is better only because it may reduce the probability for colony failure.

Genetic diversity within a colony poses two problems for bee breeding. First, a colony with a genetically diverse population of worker bees may mask the expression of genetic characteristics that are present at low frequencies. Thus colony characteristics that occur at a low frequency (such as resistance to varroa mites) may not be detectable. Second, a queen produced from a colony with a multiply mated queen is not a good genetic representative of the colony. She will have only about 0.25 (half sister) relatedness to the worker bees in that colony. Thus, it is possible that a heritable characteristic that exists in a colony will not be present in a daughter queen. In contrast, a queen produced from a single-drone mating will have 0.75 relatedness to the worker bees in the colony. Therefore, the use of single-drone mating makes it easier to both find and then retain heritable characteristics in bees.

Unfortunately, the productive life of a queen inseminated with semen from a single drone is shorter than that of a queen that is inseminated with more semen. Queens mated to a single drone usually survive less than one year (Mackensen 1964, Camargo & Gonçalves 1971). Therefore, the period of evaluation and propagation needs to be brief. Colony evaluations are traditionally conducted for an entire beekeeping season, and such projects do not use queens inseminated with a single drone because most queens would not survive the duration of the test. However, a field test that lasts only

10-14 weeks (Harbo 1996) provides an opportunity to test colonies with queens mated to single drones.

The purpose of this paper was to show that selective breeding of honey bees can be more effective when test colonies consist of only one subfamily of worker bees (colonies with queens mated to one drone). A second purpose was to define the conditions wherein singly mated queens can be used successfully in field testing.

Materials and methods

The experiment consisted of 24 colonies that were established with uniform packages of bees and a test queen. The test queens consisted of two groups of sister queens that were randomly inseminated with semen from either one or six drones. The queens began laying about 10 April and the colonies were evaluated for populations of bees and mites on 27 June.

Bees and mites for the test were collected from normal colonies of bees into a single large cage on 29 March. The following day, they were subdivided into 24 smaller cages that were small versions of commercial packages of bees. Each package with 375 ± 31 g (mean ± SD) of bees was placed in a hive with 5 combs and a caged virgin queen. To minimize drift, screens confined the bees to their colonies until after dark on 31 March. Queens were given 3 minutes of CO_2 narcosis on 4 April just before they were released from their cages and on 5 April while they were inseminated. To prevent queens from leaving their hive to mate, one wing was clipped and queen excluders were placed over the entrances.

Queens were randomly assigned a treatment (one or six drones) and then inseminated with drones that had been collected from the entrances of >40 colonies at three different apiaries. The drones were mixed and recollected into cages so that each drone

Table 1. Analysis of variance of the percent nonreproduction of mites in brood cells. Independent variables were stock (queens were from 2 different sources) and insemination (queens were inseminated with either 1 or 6 drones). Data were log transformed because of unequal variances and skewed distributions.

	df	mean square	F	P > F
Stock	1	0.002	0.12	0.77
Insemination	1	0.11	8.1	0.01
Stock * Insemination	1	0.002	0.15	0.71
Error	20	0.013		

Harbo: Value of single-drone inseminations in selective breeding

Table 2. Means ± SD of the four variables measured in the 24 colonies that began on 29 March with 375 ± 31 g of bees and 90 mites. The effect of a one- or six-drone insemination was the same in both stock types, so the 4 combinations of insemination and stock type are not listed .

Variable	1 drone(n = 13)	6 drones(n = 11)	Stock A(n = 11)	Stock B(n = 13)
Nonreproduction[1]	0.22 ± 0.16	0.06 ± 0.07	0.16 ± 0.17	0.14 ± 0.14
Cells of brood (26 May)	5286 ± 1515	5665 ± 934	5667 ± 1173	5285 ± 1369
Weight of bees (g) (27 June)	1217 ± 467	1212 ± 280	1301 ± 397	1142 ± 373
Mite population (27 June)	428 ± 168	565 ± 254	503 ± 203	481 ± 239

[1] The analysis of this variable is described in Table 1. None of the other variables showed any statistical differences.

selected for insemination was a random selection from the population that had been collected. Queens began laying about 10 April when they were 18 days old.

Colonies were evaluated for queen performance by measuring the amount of capped brood on 26 May and the weight of the bee populations on 27 June. Capped brood was measured with a wire grid with 1 inch (2.54 cm) squares. The weight of the bees in each colony was estimated by screening the entrances of all the colonies after dark on 26 June and then weighing the colonies with and without bees on the following morning (Harbo 1986).

Mite populations were measured on 27 June. The entire mite population in each colony was on the adult bees on 27 June because the queens had all been caged on 6 June, leaving no brood in the colonies. While the hive parts and frames were being weighed without bees (as mentioned in the paragraph above), a sample of ca. 150 grams of bees was taken from the population of bees that had been brushed from the combs into an empty hive body that was temporarily placed at the normal location of the colony. Thus each sample was taken from its colony as the total weight of bees was being estimated. The total number of mites in each colony could then be calculated by knowing the weights of the bees in the colony, the weight of the bees in the sample, and the number of mites in the sample.

Nonreproduction of mites in brood cells was

evaluated by counting 300 cells of brood in the purple-eyed pupal stage. 10.0 ± 4.3 (mean ± SD) cells with mites were evaluated in each colony.

In a second test, a group of 18 queens was evaluated when they were 7–10 months old. Ten queens were inseminated with 1 drone and 8 with 6 drones. The queens were reared in May, inseminated in June, and allowed to lay eggs in small colonies for the rest of the season. Colonies were moved to a remote apiary on 29 December and the number of bees in each colony was calculated on 5 January by weighing the bees and counting a subsample of bees from each colony (as described above). Capped brood was measured on 23 February and 6 March and the bee populations were estimated again on 7 March.

Data were analyzed with general linear model analysis of variance. The treatment variables were insemination (one or six drones) and stock type (queens were from two different stocks). Four analyses were conducted with each of the following serving as the dependent variable: (1) nonreproduction of mites in brood cells, (2) change in mite population, (3) change in bee population, and (4) amount of capped brood. The second experiment evaluated older queens and only the last two variables.

Data of percent nonreproduction of mites were skewed, so nonparametric statistics were used to determine if the variability of nonreproduction of mites was different within each of the two groups of colonies (colonies with 1 vs colonies with 6 sub-

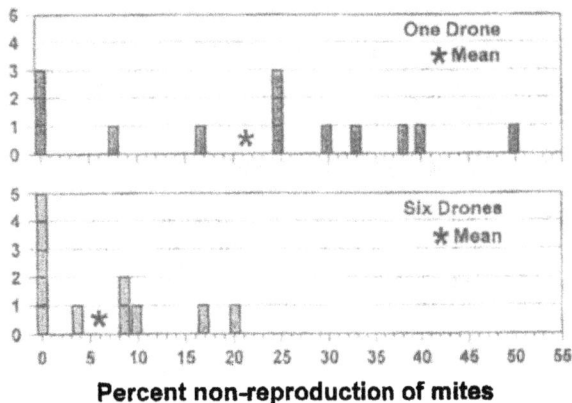

Percent non-reproduction of mites

Figure 1. Comparing the phenotypic variation among 24 colonies with queens mated to one or six drones. Each rectangle represents the percentage of varroa mites in a colony that did not produce progeny while in a brood cell. Numbers at the left indicate the number of colonies that have the same frequency of nonreproducing mites.

families of bees (data in Figure 1). Stock type was not analyzed because it had no significant effect on nonreproduction in this model (Table 1). Observations were divided into 3 classes (0-15, 16-30, and >30% nonreproduction). The 2 x 3 contingency table was analyzed with Fisher's exact test to determine if the distributions of the observations in the two groups were unequal.

Results and discussion

The single drone insemination made a marked difference in the ability to detect nonreproduction of mites in brood cells (Table 1). The other 3 variables did not show significant effects from different inseminations or different stocks of bees (Table 2). I conclude that single-drone inseminations are valuable and perhaps necessary for detecting the range of variation that can exist when measuring certain characteristics of bees at the colony level.

Comparing the distribution of the variable nonreproduction of mites in brood cells is probably more important than comparing the means (Figure 1). Based on Fisher's exact test, there was a 0.03 probability that the two groups had equal variability for nonreproduction of mites. Therefore, I conclude that the distributions were not equal and that insemination with one drone probably caused an increase in the variability among colonies.

Although it is not surprising that there was

greater variability among colonies with queens mated to single drones than among colonies with queens mated to six drones, I expected the means to be equal. The means were not equal (Table 2, Figure 1). The geometric means (perhaps more valid for these data because of the skewed distribution) were also quite different (11.8 and 2.8%). In these results it appears that when most of the bees in a colony did not possess the characteristic for nonreproducing mites, the expression of the characteristic was repressed.

Results from the first test suggest that queens mated to single drones can be used in field testing if the queens are young. When queens were less than four months old, colonies with queens mated to single drones produced as many progeny as colonies with mated to six drones (Table 2).

The age limit for using singly mated queens in field tests is somewhere between 4 and 7 months. Thus a field test needs to be short if singly mated queens are to be used. The second test was conducted with older queens, and it suggested that queens mated to single drones may not be acceptable for field testing if the queens are over 7 months old. In that test, the bee population in ten colonies with queens mated with one drone grew by a factor of 1.36 from 5 January to 7 March, whereas the population in eight colonies with queens mated to six drones grew by a factor of 2.06 (F = 11.1, P = 0.005). Thus, the groups were different, and the colonies with singly mated queens were unsatisfactory.

In most cases, queens should be inseminated and laying before they are put into uniform colonies for field testing. I did not do that in the first experiment because I was testing an insemination procedure and did not want to show bias toward either treatment by subjectively choosing queens for the test. Fortunately only three of the original 27 queens did not survive (two inseminated with one drone and one inseminated with six drones). However, in most cases the insemination procedure is not being tested, and in those cases it is helpful to cull the poor queens (for example those that have been injured and those that are not laying well).

In conclusion, colonies containing only one subfamily of bees (with queens inseminated with semen from a single drone) are valuable and perhaps necessary at times for detecting the range of variation that can exist among colonies of honey bees. This study showed that insemination with a single drone was important for detecting nonreproduction of mites in brood cells, but it may also be important for detecting other characteristics that are present in a population at low frequencies. It appears that Rothenbuhler (1960) was correct in stating that when there are many subfamilies of worker bees in a colony (a colony with a multiply mated queen), the social environmental factors imposed by bees of other subfamilies can modify the full expression of a characteristic that may appear in only one of the subfamilies.

Acknowledgment

Deborah Boykin (USDA-ARS, Stoneville, MS) provided advice on statistical analyses.

References cited

Camargo, J. M. F. and L. S. Gonçalves. 1971. Manipulation procedures in the technique of instrumental insemination of the queen honeybee *Apis mellifera* L. (Hymenoptera: Apidae). *Apidologie* 2: 239-246.

Fuchs, S. and V. Schade. 1994. Lower performance in honeybee colonies of uniform paternity. *Apidologie* 25: 155-168.

Harbo, J. R. 1986. The effect of population size on brood production, worker survival and honey gain in colonies of honeybees. *J. Apic. Res.* 25: 22-29.

Harbo, J. R. 1996. Evaluating colonies of honey bees for resistance of *Varroa jacobsoni. BeeScience* 4: 100-105.

Mackensen, O. 1964. Relation of semen volume to success in artificial insemination of queen honey bees. *J. Econ. Entomol.* 57: 581-583.

Mackensen, O. and W. C. Roberts. 1948. A Manual for the Artificial Insemination of Queen Bees. USDA, Bureau of Entomology and Plant Quarantine, ET-250. 33 pp.

Oldroyd, B. P., T. E. Rinderer, J. R. Harbo, and S. M. Buco. 1992. Effects of intracolonial genetic diversity on honey bee (Hymenoptera: Apidae) colony performance. *Ann. Entomol. Soc. Am.* 85: 335-343.

Page, R. E., G. E. Robinson, K. M. Fondryk, and M. E. Nasr. 1995. Effects of worker genotypic diversity on honey bee colony development and behavior (*Apis mellifera* L.) *Behav. Ecol. Sociobiol.* 36: 387-396.

Rothenbuhler, W. C. 1960. A technique for studying genetics of colony behavior in honey bees. *Am. Bee J.* 100:176, 198.

Key words: insemination, selection, *Varroa jacobsoni, Apis mellifera*

Genomic mapping of honey bee defensive behavior

GREG J. HUNT[1], ERNESTO GUZMÁN-NOVOA[2] AND ROBERT E. PAGE, JR.[3]

[1] DEPT. OF ENTOMOLOGY, PURDUE UNIV.,
WEST LAFAYETTE, IN 47907-1158, USA

[2] INIFAP, MEXICO CITY,
EDO. DE MEX., MEXICO

[3] DEPT. OF ENTOMOLOGY, UNIV. OF CALIFORNIA,
DAVIS, CA 95616, USA

Summary

People are beginning to use genetic mapping to identify genes in crops so that the information can be used to breed more productive plant varieties. The reason for mapping genes that affect important traits is to use them as diagnostic tools that speed up breeding. If you can find genetic markers for genes that influence those traits, you can use the markers to help select for those traits. We reasoned that the same techniques could be applied to honey bees to locate genes that affect stinging behavior. To do this, we made a cross between a gentle European queen and a drone from a highly defensive Africanized colony to get a single hybrid queen. We obtained many drones from that hybrid queen and each one was crossed to a single European queen. Each drone was the father of a separate colony. Then, we were able to evaluate the stinging behavior of those colonies that were descended from the hybrid queen and to compare specific DNA markers in each drone with the behavior of the colony that he fathered. We also evaluated the colonies for hygienic behavior. We made a map of the honey bee chromosomes with genetic markers that were generated in the polymerase chain reaction (RAPD markers). We also identified the map locations of possible genes for stinging behavior. At least one of these map locations (we call it sting-1) had an effect that was strong enough that we can be 95% sure there is a gene (or genes) nearby that affects defensive behavior. Perhaps this information will help us to breed gentle bees by using markers as diagnostic tools to test for the presence of specific genes. These same techniques could be used to study resistance of bees to varroa mites or diseases. For example, our preliminary data on hygienic behavior suggests that a gene influencing this disease-resistance trait is linked to sting-1.

History

Africanized bees are notorious for their stinging behavior because they are much more likely to sting, and to pursue intruders in large numbers (Collins and Kubasek 1982, Stort and Gonçalves 1991). Now that they are in the lower United States, people are naturally concerned about the spread of these bees and their effect on our queen industry. In Mexico, the Africanized bees have spread throughout the country and have resulted in reduced honey yields, increased labor costs and a number of deaths of both livestock and people. In the United States, Africanized bees have resulted in several deaths, and also in cumbersome state regulations of the beekeeping industry. We are interested in the possibility of finding diagnostic markers to use as tools to certify that bees do not have specific genes that influence stinging. As a first step, we would like to map the locations of the genes on the honey bee chromosomes.

Honey bees are "willing" subjects for genetic research because they have many attributes that make them easy to study. They have large colonies with many individuals, so we can test lots of them. The drones are haploid, so they only have one set of chromosomes instead of the usual two sets that most animals have. It is also possible to obtain colonies in which all of the workers have the same father (using instrumental insemination) and the father transmits the exact same set of chromosomes to each of his several thousand worker progeny. This allows us to test the effect of the drone father's genes on the behavior of the workers. In effect, the genetic similarity within a colony (due to single-drone insemination) makes it easier to see behavioral differences between colonies that are caused by the genes of different fathers.

To learn more about the genetics of stinging behavior, we used genetic markers to try to map the locations of the genes involved. These techniques have been used for several years in the breeding of crop plants (Tanksley 1993). The idea behind these methods is that it is possible to find a DNA marker (seen as a band on an agarose gel) that comes from a spot on a chromosome near a gene that affects the trait you are interested in. For example, it is possible to follow the inheritance of a marker from the African parent and to see if colonies that inherit that marker are more likely to sting. The type of marker that we used is generated in the polymerase chain reaction and the markers are called *random amplified polymorphic DNA, or RAPD markers.* These markers are very useful because you do not need much DNA to generate the markers and insects are fairly small and don't contain much DNA. In fact, we can generate thousands of markers from a single drone with this technique.

Once you have a marker for the gene, you can use it to follow the inheritance of the gene and to select for it. Some plant breeding companies are investing heavily in this technique. We believe that honey bees are as important as any one crop plant because of their pollination services, so we should be using the latest technology for honey bee breeding. Finding markers for genes may help us to make sure that we have the genes that we want in our bees. These markers could serve as another tool to use for breeding bees that are gentle and resistant to mites. Another benefit we see from this research is that it teaches us more about the effects of specific genes on honey bee behavior. The more we learn more about the actions of these genes on stinging behavior, the better our chances of being able to manipulate the bee's behavior to suit our needs.

Methods and materials

Sources of bees and crosses. We selected very gentle stocks of European honey bees from a commercial operation in Mexico and made crosses between them by single-drone insemination to establish our European population. We also produced very defensive Africanized stocks by collecting feral swarms, testing for defensive behavior, and making crosses. The Africanized queens were crossed to semen pooled

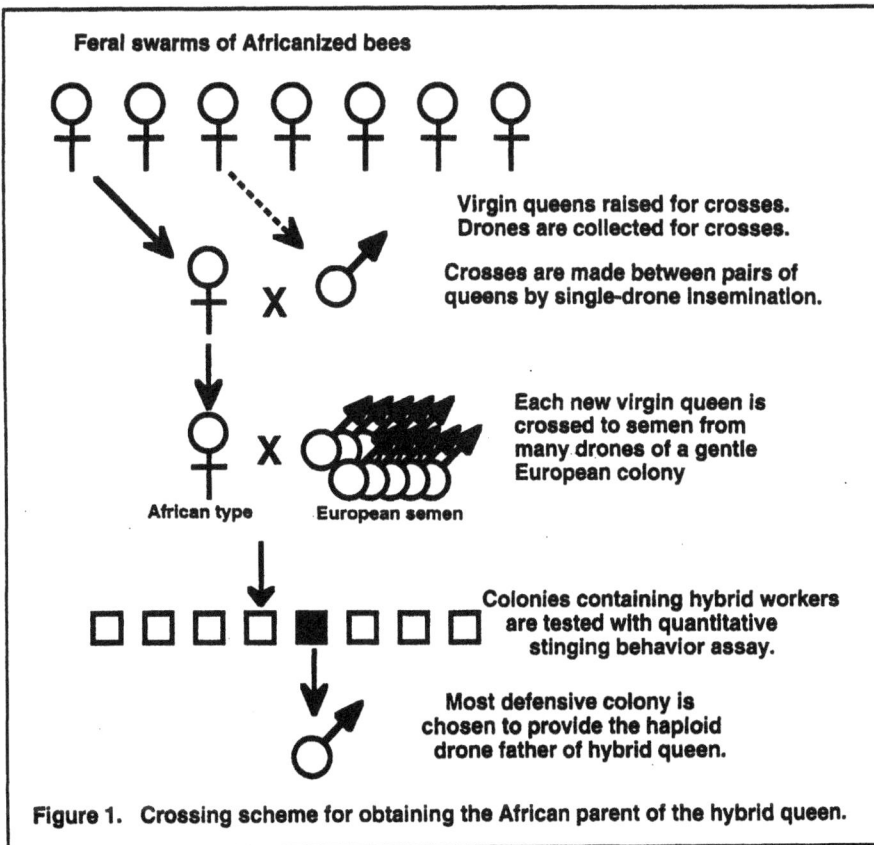

Feral swarms of Africanized bees

Virgin queens raised for crosses. Drones are collected for crosses.

Crosses are made between pairs of queens by single-drone insemination.

Each new virgin queen is crossed to semen from many drones of a gentle European colony

African type   European semen

Colonies containing hybrid workers are tested with quantitative stinging behavior assay.

Most defensive colony is chosen to provide the haploid drone father of hybrid queen.

Figure 1.   Crossing scheme for obtaining the African parent of the hybrid queen.

from a number of European drones from selected gentle stock. This cross provided colonies containing hybrid workers to test for defensive behavior so that we could choose which Africanized queen to be the parent of the hybrid queen (see Figure 1). The queen with the most defensive hybrid workers was used to provide the haploid drone father of the hybrid queen. In addition to tests of defensive behavior, the bees were analyzed morphometrically and with mitochondrial DNA testing to insure the parents of the hybrid queen had either African or European characteristics.

From the hybrid queen we obtained hundreds of drones which were kept in a special cage within the colony until they were mature. We crossed drones that from the hybrid queen individually to queens from a single European colony that was preselected as gentle. These European honey bee test queens were about 75% related to each other because they had the same haploid father. Therefore, most of the variation in behavior between the colonies should be caused by the contribution of genes from the different drone fathers. Of 313 European sister queens that were inseminated, about 250 were introduced into separate colonies. After the death or supercedure of some of the queens, 172 colonies that contained backcross workers remained and were used as our test population.

Mapping behavioral genes. Our procedure for mapping defensive behavior genes relied on techniques for mapping quantitative trait loci (or QTLs). The reasoning behind QTL mapping is pretty simple. For genes affecting defensive behavior, the drone could transmit either the African version of the gene (the African allele) or the European version. For nearby DNA markers (we used RAPD markers), the drone is likely to transmit the African version of the RAPD marker if he transmitted the African version of the nearby defensive behavior gene. So, the colonies were tested

for defensive behavior and their drone fathers were analyzed for RAPD markers. Then, we checked to see whether they had the African or European version of the marker and how this correlated with their stinging behavior. This allowed us to identify possible chromosomal regions affecting defensive behavior. The strategy for mapping defensive behavior genes is shown in Figure 2. The same type of procedure was used to try to map a gene for hygienic behavior. The computer program that was used for mapping behavioral genes was MapQTL which maps quantitative trait loci, or QTLs (Maliepaard and Van Ooijen 1994). This program uses interval mapping techniques to estimate the likelihood that a gene (or QTL) affecting stinging-behavior is present at any point on the map. Interval mapping is a kind of analysis of variance that, in our test, looks for association between African markers and the numbers of stings in our assay (Lander and Botstein 1989). We also looked for associations between markers and scores for behavior during hive manipulations. These were done with non-parametric statistical tests.

Constructing the honey bee map. We used the drones of the hybrid queen to make a map of the honey bee chromosomes. Because of male haploidy, the drones of the hybrid queen only had one set of chromosomes, so they also had only one type of marker from any point on a chromosome. Male

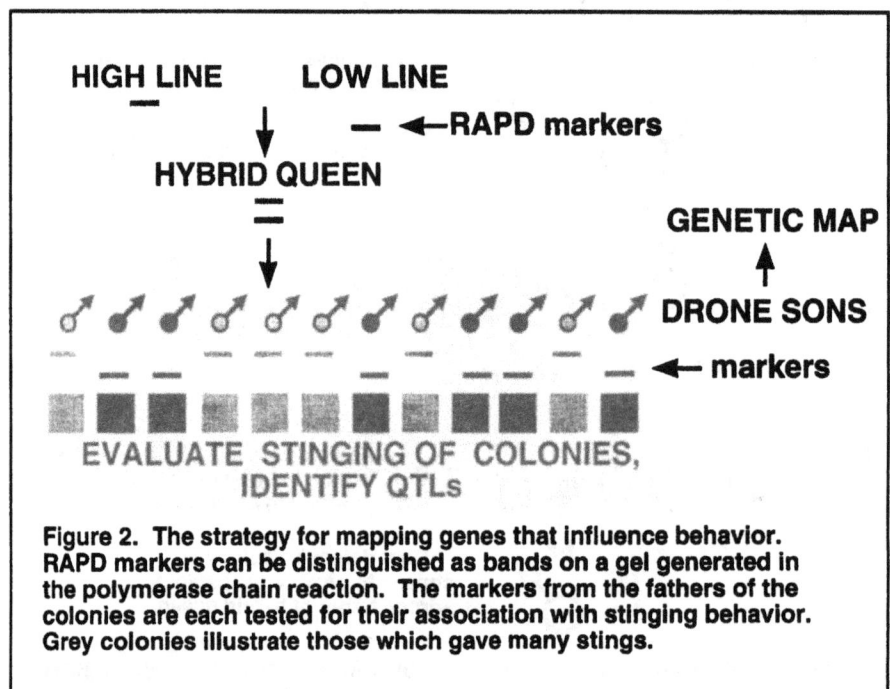

Figure 2. The strategy for mapping genes that influence behavior. RAPD markers can be distinguished as bands on a gel generated in the polymerase chain reaction. The markers from the fathers of the colonies are each tested for their association with stinging behavior. Grey colonies illustrate those which gave many stings.

haploidy simplified analyses, so we looked at the drones' markers to determine linkage relationships between the RAPD markers. Over 400 different RAPD markers (from 400 different chromosomal locations) were scored in the drones. For each drone, we could tell whether they inherited the marker from the European or Africanized parent. By determining which markers were inherited together in the drones, we could tell which markers were linked on the chromosomes and make a map of the marker positions. Mapping honey bee chromosomes was done with the same techniques that were used before (Hunt and Page 1995). A manual that explains all these techniques in detail is available (Hunt 1997a,b).

Defensive behavior assay. We had 172 test colonies containing backcross workers. Test colonies were maintained in standard Langstroth hives consisting of one deep brood box and a super for honey. Two weeks before the test, the colonies were equalized for numbers of workers by removing bees and brood from the larger colonies. One hundred and sixty-two of these test colonies were assayed for their tendency to sting (Guzmán-Novoa and Page 1993). In this test, a black suede leather patch attached to the end of a one meter white stick was moved back and forth by hand in a rhythmic way, right in the colony entrance. The time required for the first sting to occur was recorded. The bees were allowed to continue stinging the patch for one minute after the first sting. If the bees did not sting within two minutes, the test was discontinued for that colony. All of the colonies in each of the 9 apiaries were tested simultaneously to minimize behavioral interference between colonies; for example, bees of one colony reacting to alarm pheromones produced by a nearby colony. The colonies were tested four times.

Each test colony was also evaluated twice for other behavioral traits that differ between European and Africanized strains of bees. Each colony was scored for various behavioral tendencies by one researcher (Guzmán-Novoa) and his assistant. These scores were: 1) the tendency of the workers to sting during manipulations, 2) tendency to hang from combs, 3) tendency to fly up during colony manipulations. High values for all these traits are characteristic of Africanized bees. The top lid of the hive was carefully opened and 2 puffs of smoke were given to the top bars of the frames. Then, the super box was removed and 4 puffs of smoke were given

to the tops of the brood frames. Four frames of brood were removed, one at a time, and inspected. The colony was then scored on a 1-5 scale for tendency to sting, hang or fly.

Scores for stinging behavior were based on an actual a count of the average number of times the bees stung the assistant's hands during the two hive opening events (1=0 stings, 2=1 sting, 3=2-3 stings, 4=5-10 stings, 5=10-20 stings. The range of average ratings for stinging was 1 to 4.5. "Hanging" scores were based on the approximate proportion of bees hanging from the comb (1=20%, 2=40%, 3=60%, 4=80%, 5=100%). "Flying" was scored on a relative scale from 1 to 5 based on the researcher's experience.

Hygienic behavior assay. Hygienic behavior was tested by a pin-prick assay. One pupa in a sealed cell, and the six pupae around it were all pierced with a pin. This was repeated in 4 or 5 locations on a brood comb and the comb was placed back in the hive. After 48 hours, we checked to see what proportion of pupae that were killed were removed by the bees. This data was used to try to map QTLs that affect hygienic behavior.

Results

The test colonies contained workers that were about three-quarters European and most of them were not highly defensive. About half of the colonies never stung the patch, or stung less than 5 times per minute, on average (Figure 3). But a few colonies stung 150 times per minute and stung the beekeeper's hands 5-20 times during the brief hive inspection. By looking for associations between genetic markers and numbers of stings, one QTL on linkage group IV was identified that affected stinging behavior. This QTL met the threshold for controlling the genome-wide error rate for alpha = 0.05 (sting-1, Figure 4). The statistical test score (LOD score) was 3.57. This value would only occur by chance in the honey bee genome one time in twenty, so we are 95% sure that there is a gene for stinging behavior on this part of the chromosome. The estimated additive effect of an African-type gene (allele) at sting-1 was to increase the stings in the patch by 45. This region also had an effect on the tendency of the bees to sting the beekeeper (p<0.01), and the tendency for bees to fly out when the colony was opened (p<0.05). There also were 4 other possible QTLs affecting numbers of stings with lower LOD scores (near 1.5). These may indicate the positions of other genes that affect stinging behavior. Several of these QTLs, or genes were signigicant for stinging the beekeeper. One of these was very significant for stinging the beekeeper (p<0.005) and flying up during colony manipulations (p<0.0001).

The chromosomal region containing sting-1 may also be affecting hygienic behavior, but the LOD score was much lower (1.44). It is interesting to note, however, that the same marker (stsN4-.245) had the highest association with both types of behavior. In the case of hygienic behavior, a non-parametric test (Mann-Whitney) was significant at p=0.02. In our study, the tendency to sting was always inherited from the Africanized parent (associated with African-type genetic markers). However, the tendency to show good hygienic behavior came from the European parent.

Discussion

RAPD markers were very reliable for mapping honey bee chromosomes. We were able to identify the same linkage groups (maps of honey bee chromosomes) that we found before. This was possible by seeing RAPD marker-fragments that were about the same size and generated from the same ten-nucleotide DNA primer. There were more than 4 such markers in common on all the major linkage groups we looked at. Comparing our different honey bee maps is helping us to find more markers near the genes that affect stinging behavior by identifying markers that were used in the other maps.

Honey bee defensive behavior involves both guard bees and responders that fly out and sting the intruder. Our study looked at the act of stinging. We can be pretty confident that we have identified at least one gene that affects stinging behavior. But all of these QTLs (genes) need to be confirmed, preferably by an independent lab. We need more tests that show these are markers of African DNA near genes that influence stinging behavior. If we obtain markers that closely flank the genes that influence defensive behavior, they may be valuable to certify that stocks do not carry these genes in breeding programs. Studies also are needed to determine if the same genes influence stinging behavior in crosses between European strains of honey bees in the U.S. Our program now is designed to confirm the effects of these QTLs on the behavior of individual bees and to use different behavioral assays to see if the different genes are affecting behavior in different ways. These assays will compare the RAPD markers of individual workers with their behavior. For example, we will sample bees that are found to be the first to sting the patch, bees that guard, and bees that respond to alarm pheromone.

Hygienic behavior is important for a colony's ability to overcome disease and varroa mites (Rothenbuhler 1964a,b; Gilliam et al. 1988; Boeking and Drescher 1992; Spivak 1996). After searching all of the honey bee chromosomes, we only found one small LOD score peak for hygienic behavior as evidence for a nearby gene affecting this important behavior. But this peak corresponded with our biggest peak for stinging behavior, suggesting that the same gene that affects stinging, or a nearby gene, affects hygienic behavior. Our results on hygienic behavior are very preliminary. We would like to confirm the effects with a freeze-killed brood assay.

The classic work of Dr.

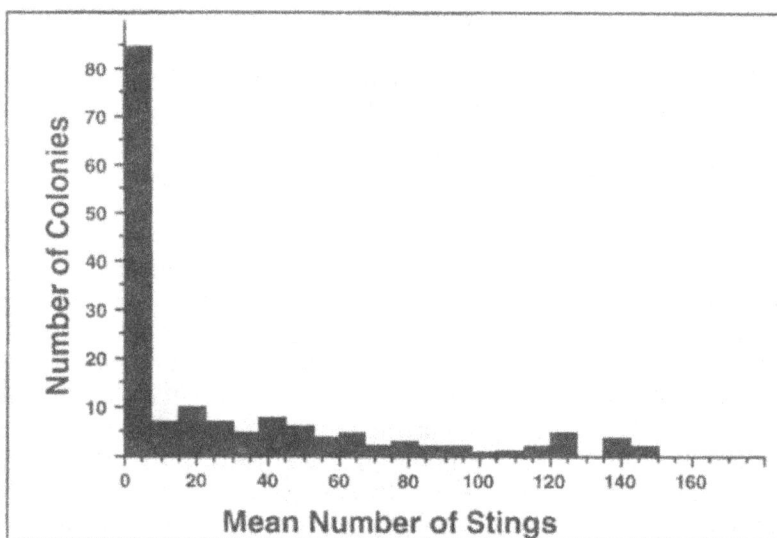

Figure 3. Average number of stings observed in one minute during the four trials of the defensive-behavior assay for 162 colonies.

Hunt et al. Genomic mapping of defensive behavior

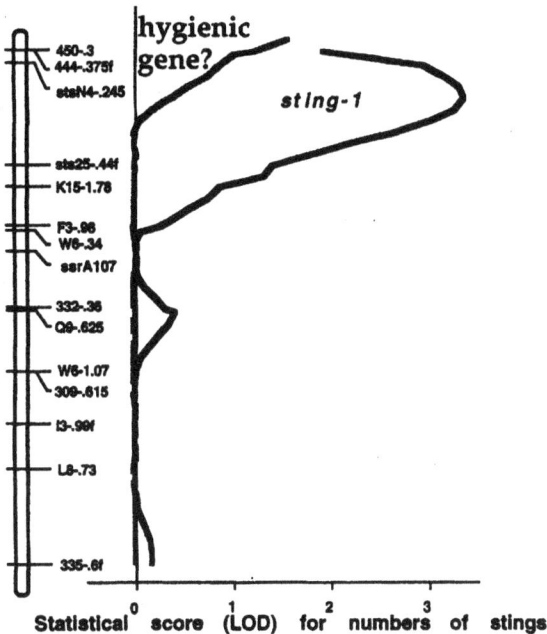

Figure 4. Linkage froup IV, corresponding to one of the 16 chromosomes of the honey bee and the likelihoods that a quantitative trait locus (gene) influencing two behaviors exists at any point on the chromosome. The black bar incicates a high probability for a stinging-behavior gene; the grey bar indicates the probability for a hygienic-behavior gene at about the same location. The liklihood score (LOD) was 3.6 for sting-1.

Rothenbuhler on honey bee hygienic behavior involved two lines of bees, one line was very defensive and hygienic, and the other was non-hygienic and gentle. It is also possible to select for honey bees that are both hygienic and gentle. In our study, the tendency to be hygienic seemed to come from the gentle European parent but there was evidence that the genes affecting stinging and hygienic behavior may be linked on the chromosome. If this turns out to be the case, it should be fairly easy to get hygienic bees that are gentle and to maintain these two traits in the same stock. If we know that the major stinging behavior and hygienic behavior genes are linked, we could then maintain good stocks by selecting for one or the other of the two traits, and the other trait would be carried along with it because the genes are linked.

References

Collins, A. M. and K. J. Kubasek. 1982. Field test of honey bee (Hymenoptera: Apidae) colony defensive behavior. *Ann Entomol Soc Am* 75: 383-387.

Gilliam, M., S. Taber III, B. J. Lorenz and D. B. Prest, 1988 Factors affecting development of chalkbrood disease in colonies of honey bees, *Apis mellifera*, fed pollen contaminated with *Ascosphaera apis*. *J. Invert. Path.* 52:314-325.

Guzmán-Novoa, E. and R. E. Page, Jr. 1993. Backcrossing Africanized honey bee queens to European drones reduces colony defensive behavior. *Ann Entomol Soc Am* 86:352-355.

Hunt, G. J. 1997a Construction of linkage maps with RAPD markers. pp 187-200 *In*: Fingerprinting methods based on arbitrarily primed PCR. M. R. Micheli and R. Bova, eds., Springer-Verlag, Berlin.

Hunt, G. J. 1997b Insect DNA extraction protocol. pp 21-24 *In*: Fingerprinting methods based on arbitrarily primed PCR. M. R. Micheli and R. Bova, eds., Springer-Verlag, Berlin.

Hunt, G. J. and R. E. Page, Jr. 1995 Linkage map of the honey bee, Apis mellifera , based on RAPD markers. *Genetics* 139:1371-1382.

Lander, E. S. and D. Botstein. 1989. Mapping Mendelian factors underlying quantitative traits. *Genetics* 121: 185-199.

Maliepaard, C. and J.W. Van Ooijen 1994 QTL mapping in a full-sib family of an outcrossing species. pp. 140-146 *In*: Van Ooijen, J. W. and J. Jansen (Eds.) Biometrics in Plant Breeding:Applications of Molecular Markers, Proceedings of the Ninth Meeting of the EUCARPIA Drvyion Biometrics in Plant Breeding.

Rothernbuhler, W. C. 1964a Behavior genetics of nest cleaning in honey bees. I. Responses of four inbred lines to disease-killed brood. *Animal Behavior* 12:578-583.

Rothernbuhler, W. C. 1964b Behavior genetics of nest cleaning in honey bees. IV. Responses of F$_1$ and back-cross generations to disease-killed brood. *American Zoologist* 4:111-123.

Spivak, M. 1996 Honey bee hygienic behavior and defense against *Varroa jacobsoni. Apidologie* 27:245-260.

Stort, A C and L S Gonçalves. 1991. Genetics of defensive behavior II. pp. 329-356 in: The "African" Honey Bee. San Francisco, Westview Press.

Tanksley, S. D. 1993 Mapping polygenes. *Annual Review of Genetics* 27: 205-233.

Keywords: Defensive behavior, aggression, *Apis mellifera*, quantitative trait loci, hygienic behavior, Africanized bees.

Breeding honey bees in Africanized areas

ERNESTO GUZMÁN-NOVOA[1] AND ROBERT E. PAGE JR.[2]

[1] CENIFMA-INIFAP, SANTA CRÚZ # 29-B,
LAS HACIENDAS, 52140 METEPÉC, EDO DE MÉX, MEXICO

[2] DEPARTMENT OF ENTOMOLOGY,
UNIVERSITY OF CALIFORNIA, DAVIS, CA 95616, USA

Introduction

African bees (*Apis mellifera scutellata*) were introduced into Brazil in 1956 as part of a selective breeding program designed to produce a bee that was better adapted to tropical conditions (Kerr 1967). African bees became feral and interbred with the local populations of European bees, producing the Africanized honey bees. These bees have since spread through most of the Americas, and reached Mexico in 1986 (Moffett *et al.* 1987).

Africanized bees have several characteristics that make them undesirable for beekeeping. They are significantly more defensive (Stort 1975a,b,c, Collins and Kubasek 1982, Collins *et al.* 1982, Villa 1988, Guzmán-Novoa and Page 1993, 1994a) and less productive (Rinderer 1985, Rinderer *et al.* 1985, Cobey and Locke 1986, Guzmán-Novoa 1986, 1996, Swezey 1986, Hellmich and Rinderer 1991, Rinderer and Collins 1991, Guzmán-Novoa and Page 1994b) than European bees. Additionally, due to the relocation of apiaries, more labor invested, purchase of better protective equipment, and a more intensive requeening of colonies, beekeeping with Africanized bees is more expensive than with European bees (Guzmán-Novoa and Page 1994b).

One possible alternative solution to these problems occasioned by the Africanization of commercial bee colonies, could be the development of methods for breeding productive and gentle bees. Here we report results of a practical breeding program conducted in an Africanized area, with the objective of keeping productive and manageable bee stocks without maintaining genetically isolated populations. The reason for not attempting to genetically isolate populations is that we believe it is necessary to develop simple systems suitable for most queen breeders. Very few breeders will be able to use instrumental insemination in their operations.

Methods

We have conducted five generations of selection in Vita-Real, a commercial operation with more than 3,000 colonies. Vita-Real is based in Ixtapan de la Sal, Mexico, which is located about 150 km southwest of Mexico City (19° N, 99° W). This area has been Africanized since 1990 (Guzmán-Novoa and Page 1994b).

The selection method consists of five steps that are yearly repeated:

1. Preselection of the most productive colonies (5-8% of all the colonies).

2. Measurement of the preselected colonies defensiveness, keeping the 25% least defensive.

3. Assessment of the capped brood pattern, and sampling the worker population of the least defensive colonies.

4. Determination of the average worker's forewing length, from the sampled colonies with the most uniform brood patterns.

5. Selection of the queens heading the colonies having workers with average forewing lengths of at

least 9.1 mm (about 15% of the preselected colonies).

Honey production is measured for each colony by counting the numbers of combs of honey harvested. Data for each colony are entered into a computer file and evaluated by Statview® (Abacus Concepts Inc.), a statistical analysis program. Each colony is evaluated against the average production of its own apiary, and assigned a Z score, which is a statistical measurement of how much greater or smaller that colony's honey yield is from the apiary average (see Rinderer 1986 for a more detailed explanation on how to use Z scores in selective breeding). Colonies are then listed in descending order according to their Z score. The top 5-8% of the colonies are selected for defensive behavior assays (second step

Defensive behavior is measured with the assay described by Guzmán-Novoa and Page (1993, 1994a), and by Guzmán-Novoa et al. (1997), which consists of a black suede leather patch (10 x 8 cm) suspended on a piece of white wood (0.7 x 0.5 x 100 cm) which we call "flag." A flag is rhythmically elevated (~4 cm) and lowered (~4 cm; two swings per s) approximately 5-10 cm in front of the entrance of each hive. The bees are permitted to sting the patch during an interval of 60 s. After each trial, the leather patches are packed and sealed in marked 20 ml plastic vials, for subsequent sting counts. This assay is conducted twice in each colony. All colonies in an apiary are tested simultaneously in order to decrease the likelihood that bees from a single defensive hive sting the leather patches presented to others. About 25% least defensive colonies (with less average stings) are subjected to the procedures of the third selection step.

The next step is to check the brood pattern for the least defensive colonies. Ten workers from each of the colonies with the most uniform brood patterns are collected and subjected to forewing length measurements (FABIS I, see Sylvester and Rinderer 1987). This step was included in the selection methodology, because we have found a high and significant negative correlation between defensiveness and forewing length (Guzmán-Novoa and Page 1994b, Guzmán-Novoa et al. 1994). Small Africanized bees tend to be more defensive than large managed, commercial bees. Approximately 30 queens from superior performing colonies with average forewing lengths of at least 9.1 mm are finally selected as queen mothers.

Virgin queens produced from the selected queen mothers are mated in mating yards. The first queens produced are used to requeen the drone mother colonies. The rest of the colonies (ca. 3,000) are then requeened.

In order to assess the efficacy of the selection method described above, we established the following controls:

1. Comparison of the honey yield per colony between our selected and an unselected population of at least 780 colonies established in the same general honey producing area.

2. Comparison of parameters supposedly affected by selection, and which were annually obtained from a random sample of colonies from the selected population. These parameters were:

a). Number of stings per min in a leather flag (Guzmán-Novoa and Page 1993, 1994a, Guzmán-Novoa et al. 1997).

b). Forewing length (Sylvester and Rinderer 1987).

c). Percent colonies with bees having African and European mitochondrial DNA (Hall and Smith 1991, Nielsen et al. unpublished data).

Data were subjected to analyses of variance and correlation analyses (Sokal and Rohlf 1981).

Results

After five years of selection, honey yields per colony were stable over time, with a slight increase of 15.9%, whereas in unselected populations, productivity decreased over 34%.

The stinging behavior of colonies from the selected population decreased more than 54%, whereas the wing length of their workers increased significantly.

The percentage of colonies containing bees with African mitochondrial DNA, decreased from 27.9% before selection, to 7.5% after four generations of selection, whereas that of colonies having bees with European mitochondrial DNA increased from 60 to nearly 80%.

Honey production was correlated neither to wing length, nor to stinging behavior, but stinging behavior and forewing length had a significant negative correlation ($r = -0.54$, $n = 947$; $P < 0.0001$).

Discussion

Our results suggest that it is possible to maintain honey yields and breed manageable bees in Africanized areas. It seems that the process of Africanization can be reversed to a certain degree with a selection program like the one described here, and it is not necessary to use instrumental insemination of queen bees to achieve these results.

The fact that honey production was correlated neither to stinging behavior, nor to forewing length, suggests that there may not be genes in common for these characteristics, which is desirable, because colonies could be selected for high honey production and low defensiveness. However, the high correlation between stinging behavior and forewing length evidences an association between size and defensiveness. This association may not necessarily be a genetic one. It could be the result of the effect of the smaller size of Africanized bees that tend to be more defensive.

We are currently working on simplifying our selection techniques in order to make them available to most queen breeders. We are particularly interested on simplifying the second and fourth steps of the process.

Acknowledgments

We are grateful to Guillermo García, Vita-Real owner, who provided labor force, colonies, and laboratory space. Miguel Arechavaleta, José L. Uribe, Froylan Gutiérrez, José Calvo, Daniel Prieto, and Enrique Estrada provided valuable assistance in various ways. This study was supported by funds of INIFAP-SAGAR, and by contracts from the California Department of Food and Agriculture.

References cited

Cobey, S. and S. Locke. 1986. The Africanized bee: A tour of Central America. *Am. Bee J.* 126: 434-440.

Collins, A. M. and K. J. Kubasek. 1982. Field test of honey bee (Hymenoptera: Apidae) colony defensive behavior. *Ann. Entomol. Soc. Am.* 75: 383-387.

Collins, A. M., T. E. Rinderer, J. R. Harbo and A. B. Bolten. 1982. Colony defense by Africanized and European honey bees. *Science* 218: 72-74.

Guzmán-Novoa, E. 1986. Apicultura y abejas Africanizadas [Apiculture and Africanized Bees]. Ed. Quetzalcoatl. México, D.F. 71 pp.

Guzmán-Novoa, E. 1996. La apicultura en México y Centro América [Beekeeping in Mexico and Central America]. *In* Memorias V Congr. Iberolatin. Apic. pp. 14-17.

Guzmán-Novoa, E. and R. E. Page, Jr. 1993. Backcrossing Africanized honey bee (*Apis mellifera* L.) queens to European drones reduces colony defensive behavior. *Ann. Entomol. Soc. Am.* 86: 352-355.

Guzmán-Novoa, E. and R. E. Page, Jr. 1994a. Genetic dominance and worker interactions affect honey bee colony defense. *Behav. Ecol.* 5: 91-97.

Guzmán-Novoa, E. and R. E. Page, Jr. 1994b. The im-pact of Africanized bees on Mexican beekeeping. *Am. Bee J.* 134: 101-106.

Guzmán-Novoa, E., R. E. Page, Jr. and M. K. Fondrk. 1994. Morphometric techniques do not detect intermediate and low levels of africanization in honey bee (*Apis mellifera* L.) colonies. *Ann. Entomol. Soc. Am.* 87: 507-515.

Guzmán-Novoa, E., R. E. Page, Jr., H. G. Spangler and E. H. Erickson, Jr. 1997. A comparison of two assays to test the defensive behavior of honey bees (*Apis mellifera* L.). *Bee Sci.* In Press.

Hall, H. G. and D. R. Smith. 1991. Distinguishing African and European honeybee matrilines using amplified mitochondrial DNA. *Proc. Nat. Acad. Sci.* USA 88: 4548-4552.

Hellmich, R. L. and T. E. Rinderer. 1991. Beekeeping in Venezuela. *In* The "African" Honey Bee. Ed. M. Spivak, D. J. C. Fletcher, M. D. Breed, pp. 399-411. Westview Press, Boulder, Col.

Kerr, W. E. 1967. The history of the introduction of African bees to Brazil. *S. Afr. Bee J.* 39: 3-5.

Moffett, J. O., D. L. Maki, T. Andre and M. M. Fierro. 1987. The Africanized bee in Chiapas, Mexico. *Am. Bee J.* 127: 517-519, 525.

Rinderer, T. E. 1985. Africanized honeybees in Venezuela: Honey production and foraging behavior. *In* Apiculture in Tropical Climates, Vol. 3. Ed. M. Adey, pp. 112-116. Int. Bee Res. Assoc., Gerrards Cross, UK.

Rinderer, T. E. 1986. Selection. *In* Bee Genetics and Breeding, ed. T. E. Rinderer, pp. 305-321. Academic Press Inc., Orlando, Fla.

Rinderer, T. E. and A. M. Collins. 1991. Foraging behavior and honey production. *In* The "African" Honey Bee, ed. M. Spivak, D. J. C. Fletcher, M. D. Breed, pp. 235-257. Westview Press, Boulder, Col.

Rinderer, T. E., A. M. Collins, and K. W. Tucker. 1985. Honey production and underlying nectar harvesting activities of Africanized and European honeybees. *J. Apic. Res.* 23: 161-167.

Sokal R. R. and F. J. Rohlf. 1981. Biometry. New York.

Stort, A. C. 1975a. Genetic study of the aggressiveness of two subspecies of Apis mellifera in Brazil. II. Time at which the first sting reached the leather ball. *J. Apic. Res.* 14: 171-175.

Stort, A. C. 1975b. Genetic study of the aggressiveness of two subspecies of *Apis mellifera* in Brazil. IV. Number of stings in the gloves of the observer. *Behav. Genet.* 5: 269-274.

Stort, A. C. 1975c. Genetic study of the aggressiveness of two subspecies of *Apis mellifera* in Brazil. V. Number of stings in the leather ball. *J. Kans. Entomol. Soc.* 48: 381-387.

Swezey, S. L. 1986. Africanized honey bees arrive in Nicaragua. *Am. Bee J.* 126: 283-287.

Sylvester, H. A. and T. E. Rinderer. 1987. Fast Africanized bee identification system (FABIS) manual. *Am. Bee J.* 127: 511-516.

Villa, J. D. 1988. Defensive behaviour of Africanized and European honeybees at two elevations in Colombia. *J. Apic. Res.* 27: 141-145.

Key words: Africanized honey bees, breeding, selection, defensiveness, honey production

Foraging task organization by honey bees

JENNIFER H. FEWELL
DEPARTMENT OF BIOLOGY
ARIZONA STATE UNIVERSITY
TEMPE AZ 85287-1501

Abstract

In this chapter I discuss the hypothesis that genotypic variation plays a central role in task regulation by honey bee colonies. I focus on a specific model for the interaction between genetic variation and task performance, the stimulus threshold model, and on a specific set of behavioral tasks, pollen and nectar foraging. The basic tenet of the stimulus threshold model is that workers within a colony vary in their intrinsic thresholds for a given task. As the stimulus levels for the task move beyond that threshold, the worker begins to perform it. From this we can generate a colony model in which genotypic variation in the hive generates a set of specialists for a given task, but also allows the colony to respond flexibly to increased task need by recruitment of workers with higher thresholds. I present tests for two predictions of the diversity hypothesis: (1) as need for a task increases we expect an increase in the genotypic diversity of the workers in that task group, and (2) if we decrease the genotypic diversity within the worker population we decrease the colony's ability to respond flexibly to changes in task need. These predictions were tested by manipulating colony need for pollen, and measuring the responses of individual workers of known genotypes. The experimental results showed strong support for the model. In colonies with workers from naturally mated sources the evenness of representation of focal genotypes in the pollen foraging group increased as pollen need increased.

However, when workers from lines selected for high or low pollen foraging were used as focal groups, they did not vary foraging activity in response to similar changes in pollen need. In a second section of the chapter I discuss the potential for integration of the stimulus threshold model with behavioral models of task regulation. In particular, I examine the expectations of an integrated model combining genetically-based foraging preferences based on simple Mendelian inheritance and a behavioral mechanism of social information transfer. This combination of genetic and behavioral mechanisms can generate a fast and coordinated colony response to changes in task need. I tested this prediction by varying the level of pollen stores in hives in a graded manner. As predicted, colonies responded in a strong stepwise manner as storage levels moved above or below a specific set-point level. I argue that this type of integrated model represents a necessary next step in our analysis of social organization.

Introduction

One of the most distinctive characteristics of the honey bee colony is the way in which it divides its work force into subsets of task specialists. This division of labor is considered one of the most important adaptations of social insects, and is one of the most studied aspects of their behavior. Task regulation in social insects is not a simple set allocation of workers to tasks. Allocation of workers among tasks must be flexible, because the need for a given task can vary dramatically across short time scales. Indeed, one of the key characteristics of division of labor in social insects is that they show a remarkable ability to respond to changes in need for a given behavior (exemplified by the defensive response of African bees!).

The question of how honey bee hives can behave so flexibly at the colony level, but maintain individual worker specialization is an important one. In honey bees foraging is a particularly dynamic task system, because hives must respond both to variation in colony need and resource availability. Nectar and pollen availability can vary from a daily to a seasonal basis. Resource need, especially for pollen, also fluctuates as colony brood production increases and decreases through the year (see Winston 1987 for review). Foraging in honey bees provides an ideal context in which to address questions about task specialization and flexibility, because pollen and nectar collection are two discrete measurable tasks. Foraging behavior in honey bees is also an important topic in its own right, because pollen and nectar collection are why the majority of us keep bees.

Genetic mechanisms of task regulation in honey bees

How do colonies respond to changes in the need for a particular task, such as nectar or pollen collection? We can categorize current models of variation in task performance in social insect colonies into two approaches: those that focus on the interaction between workers and environmental cues, and those that focus on the effects of genotype on worker task choice. A majority of the current models of task performance in social insects make the underlying assumption that workers assess the environment similarly, and that variation in task performance is generated through variation in these stimulus cues (Seeley 1986, 1989; Seeley and Levien 1987; Seeley *et al.* 1991; Tofts and Franks 1992; Franks and Tofts 1994, 1996). These models have contributed greatly to our understanding of how worker interactions with the environment contribute to task choice. However, they do not integrate the growing evidence that honey bee workers vary innately in how they assess task cues, and that this variation has a potentially strong effect on colony behavior. Studies have found strong ge-

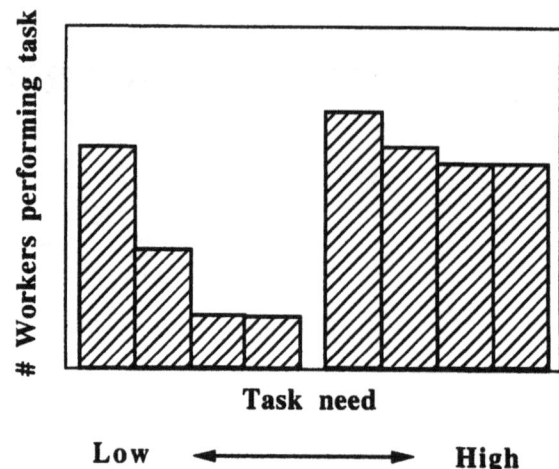

Figure 1. The expected change in the number of workers from different genetic subgroups performing a task as the stimulus level for the task increases. As task need increases we expect that the representation of workers from the different subgroups should become more evenly distributed.

netically based variation in worker performance of numerous tasks, including queen rearing, guarding the nest, removal of dead workers, and foraging behavior (Calderone and Page 1988, 1991; Frumhoff and Baker 1988; Robinson and Page 1988, 1989b). Given this, we need to address the question of whether task allocation may be regulated via variation in worker genotypes.

The principal current model considering genotypic variation as a primary factor in determining task allocation is the stimulus threshold model, also known as the fixed stimulus model (Robinson and Page 1989; Calderone and Page 1991; Page and Robinson 1991; Bonabeau *et al.* 1996). The stimulus threshold model assumes that individuals within a colony vary genetically in their sensitivity to stimuli for a given task. For each worker, when the stimulus for a task reaches their intrinsic threshold they begin performing it. When the stimulus for a task is low, a narrow genetic subset of workers performs the task. These are the specialists. As the stimulus level increases, a wider range of genetic thresholds are met and more workers begin to perform the task. Colony flexibility is generated when there is a high level of genotypic variation among the workers, because diversity produces a wide range of worker thresholds for a given task. Genotypic diversity in honey bee colonies results in part from queens mating with multiple males, generating genotypically different subfamilies of

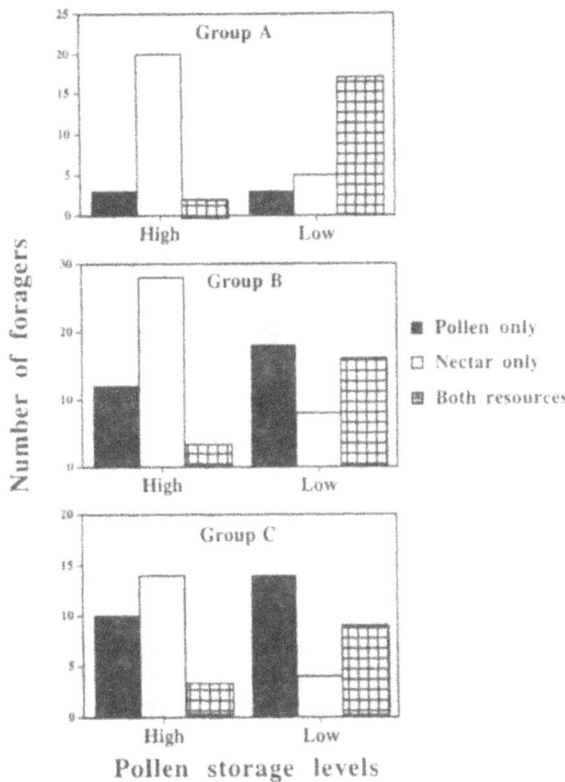

Figure 2. The number of foragers from three genetic groups (A, B and C) collecting (1) only pollen, (2) only nectar, or (3) both resources under conditions of high pollen stores (low stimulus for pollen collection), or low pollen stores (high stimulus for pollen collection). Each treatment was presented for five days. The three groups varied significantly in their distribution of workers among foraging tasks (G = 18.8; P < 0.01). Each group increased the number of foragers collecting pollen or both resources as need for pollen increased (G = 49.001). From Fewell and Page (1993).

workers in the hive (Page 1980).

This model makes two key predictions. First, we should expect that if genotypic diversity contributes to task regulation, then increased need for a task produces increased diversity among the workers performing that task. Under conditions of low need the genotype(s) most sensitive to that task would be over-represented. As we increase the need for the task, workers from other genotypic groups would begin performing the task also, increasing the diversity, or specifically the evenness of representation of genetic subfamilies in the task group (Figure 1). It is worth noting that this prediction is not consistent with the hypothesis that task regulation occurs primarily through individual behav-

Fewell: Foraging task organization by honey bees

ioral flexibility. Non-genetic models of task performance assume that task performance is random with respect to genotype, so we should not expect any genotypic change associated with the behavioral response.

A second key prediction from the model is that if genotypic variation is a central or driving factor in task flexibility, we should be able to reduce colony flexibility via a reduction in genotypic diversity. For example, in the case of foraging, if we reduce genetic diversity, the colony should not respond as well to environmental stresses that normally elicit a change in foraging behavior. Again, this is a key prediction of the stimulus threshold model, because it demonstrates a causal relationship between genotypic diversity and colony function. This relationship would be inconsistent with models of task regulation based solely on behavioral responses to environmental change.

Genotypic diversity and foraging for pollen

Honey bees show flexibility in numerous aspects of foraging behavior. However, pollen foraging is a task in which hives respond in a particularly dynamic way to changes in need. The primary function of pollen foraging is to provide nutrition to developing brood (Winston 1987). In previous work Mark Winston and I found that pollen stores in the hive are regulated homeostatically around a specific set-point (Fewell and Winston 1992), which in turn is related to brood production (Free 1967; Eckert *et al.* 1994). As pollen storage levels move below or above the set-point, hives dramatically alter their foraging behavior to adjust the amount of pollen coming into the hive (Fewell and Winston 1992, Camazine 1993). The predictability of this response makes it a useful system for testing the mechanisms of colony flexibility.

I tested the predictions of the stimulus threshold model in a series of collaborative experiments in which we examined how both individual workers and colonies as a whole respond to changes in the need for pollen. In each of these experiments we placed small hives (approximately 5000 workers) into screen mesh tents and supplied the tents with pollen and nectar resource stations (see Fewell and Page 1993 for detailed methodology). The use of screen tents allowed us to control access to pollen and nectar resources, and to monitor the foraging behavior of individual workers at the stations.

Experiment 1: Is colony response to changes in pollen need associated with changes in the genotypic composition of the foraging group?

In collaboration with Rob Page at the University of California, Davis, I tested the prediction that an increase in need for pollen would produce an increased pollen foraging, and a corresponding increase in the diversity of workers collecting pollen. To test this we placed newly emerged, individually marked workers from three different genotypic sources (Genotypic groups A, B and C) into a hive, and waited for them to begin foraging. We then manipulated pollen stores in the colony, first increasing and then decreasing storage levels. The colony's pollen foraging activity level was dramatically higher under conditions of low versus high pollen stores (Fewell and Page 1993). This response was similar to that seen previously in large field colonies (Fewell and Winston 1992) in response to similar manipulations.

The change in foraging activity was accompanied by a distinct change in the genotypes of the workers collecting pollen. Under conditions of low pollen need the three groups varied significantly in their foraging behavior. Group A and B workers showed a strong preference for nectar collection, although a slightly higher number of group B workers collected pollen. A significant number of group C workers collected pollen, even though the colony need for the resource was low (Fewell and Page 1993; Figure 2).

When pollen stores were removed the pollen foraging activity of all three groups increased. However, the increase in number of workers collecting pollen was much higher for groups A and B, which were under-represented under low pollen storage conditions, than for group C (Figure 2). The genotypic composition of the pollen foraging group moved from over-representation of one genetic subgroup (group C) to a more even distribution of the focal subgroups as need for that foraging task increased (Figure 3). These data are consistent with those of subsequent experiments both in mesh tents (Fewell and Bertram submitted) and in field colonies with naturally mated queens (Dubas and Fewell in prep) in which we see a similar response; colonies respond to increased need for a foraging task by increasing the genetic diversity of their foragers.

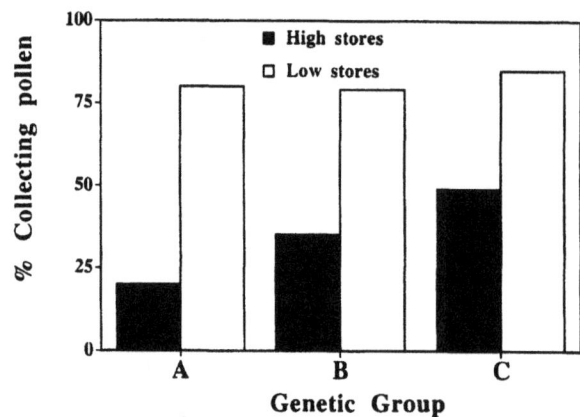

Figure 3. The proportion of foragers within each of the three genetic groups (A, B and C) that collected pollen under conditions of high pollen stores (low stimulus for pollen collection), or low pollen stores (high stimulus for pollen collection). This analysis included those foragers that collected only pollen and those that collected both resources. The groups varied in the proportion of workers collecting pollen when need for the task was low. As colony pollen need increased the proportion of workers collecting pollen increased to levels among the three groups.

Experiment 2: Does reducing genotypic diversity within a colony constrain its ability to respond flexibly to changes in pollen need?

The previous experiment showed that genotypic variation is linked to task flexibility, but it did not establish a causal relationship between genotypic diversity and foraging regulation. As part of an experiment examining inheritance patterns for pollen and nectar collection, I tested the prediction that a reduction in genotypic diversity will cause a corresponding reduction in colony foraging response to changes in pollen need. Again in collaboration with Rob Page, I manipulated the pollen storage levels in small honey bee hives placed inside mesh tents. We manipulated two hives, using a similar protocol to that of experiment 1. However, this time we added marked workers from genetic lines selected through 3 generations by Rob Page and Kim Fondrk at UC Davis for either high or low levels of pollen stores (Page and Fondrk 1996). We added workers both from the low pollen hoarding line (Group R) and the high pollen hoarding line (Group T). We also added workers from a third group (Group S) which was a hybrid of the other two.

As in experiment 1 we saw significant variation in behavior among the groups. Predictably Group

R showed a strong preference for nectar collection, while T was over-represented in the pollen foraging population. Our expectation was that Group S workers would be intermediate in their resource preference. However, this group showed a high preference for nectar foraging similar to Group R (Fewell and Page submitted; Figure 4).

The lack of an intermediate phenotype in our focal honey bees meant that only the tail ends of high and low thresholds for pollen collection were represented in the group. Thus, we had reduced both the genotypic variation and the potential variation in phenotypic expression of pollen collection within the hives. These results support the prediction that a reduction in diversity constrains the hives' foraging flexibility. In contrast to the results of experiment 1, each of the groups remained constant in their resource preferences despite a large change in the colonies' need for those resources (Figure 4). This experiment was repeated in another study after four generations of selection, with similar results (Page and Fewell in prep). In each case we see a reduction in behavioral response to changes in colony pollen need.

The finding that genotypic diversity plays a regulatory role in foraging has important practical implications. If genotypic diversity contributes to colony flexibility, then breeding practices that reduce genetic diversity in the hive can have a negative impact on colony productivity. This seems to be the case. In a field experiment examining variation in queen mating, Oldroyd *et al.* (1992) found that hives with multiply mated queens were more successful in brood production. Page and Fondrk (1996) found that hives with queens selected over three generations for high pollen intake had lower brood production than unselected hives, because pollen stores interfered with brood comb.

The inheritance mode of pollen and nectar foraging

The genotypic basis to a behavioral trait has a potentially large impact on its expression. Variation in the phenotypic expression of task thresholds (what level of stimulus actually causes the bee to perform a task) can range from being continuous to bimodal or even discrete. Continuous variation in task thresholds would produce a range of workers from those with strong sensitivities to cues for a task to those who are extremely unlikely to respond, with intermediate behavioral patterns well

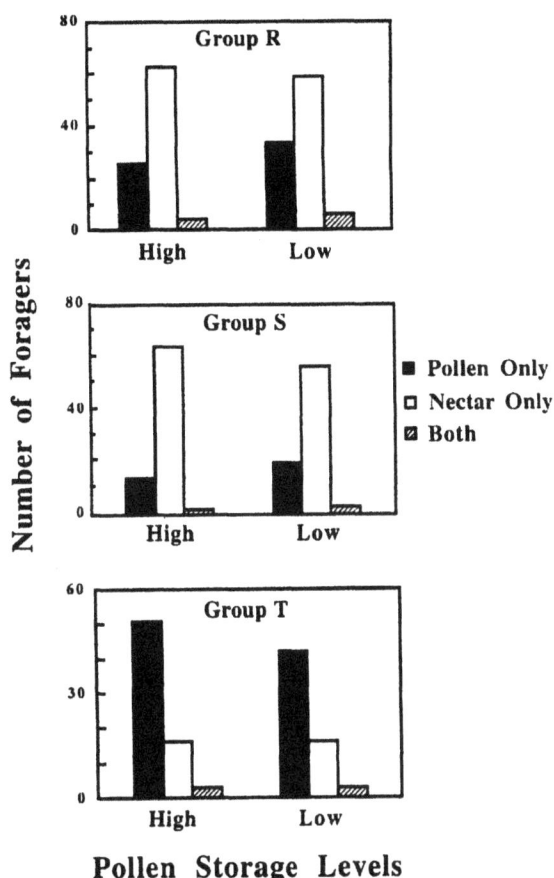

Pollen Storage Levels

Figure 4. The number of foragers from three genetic groups (R, S and T) collecting (1) only pollen, (2) only nectar, or (3) both resources under conditions of high pollen stores (low stimulus for pollen collection), or low pollen stores (high stimulus for pollen collection). Colonies were observed for three days within each treatment. Group R consisted of workers from lines selected for low pollen intake, Group T consisted of workers from lines selected for high pollen intake. Group S workers were hybrids of the two selected lines. Group T differed significantly from Groups R and S in the distribution of workers performing different foraging tasks ($X^2 = 68.5$, P = 0.000). There was no significant change in behavior as colony storage levels were increased or decreased.

represented. This could be generated from variation in the developmental environment of workers, or from thresholds being based on underlying additive genetic variation, generated when several genetic loci with multiple alleles contribute fairly equally to a trait.

The alternative to continuous variation in task preference thresholds would be a model closer to classical Mendelian inheritance, which divides traits into more discrete categories. A limited number of genes contributing to foraging task preference could reduce the distribution of worker foraging preferences towards a bimodal distribution, producing a set of pollen specialists and non-specialists. Low levels of task switching for workers already collecting pollen, (as seen across our experiments: Fewell and Page 1993; Fewell and Bertram submitted; Fewell and Page submitted), also contribute to a bimodal distribution.

The fast response to selection for foraging (after just 3 generations) indicates that variation in pollen and nectar foraging has a strong genetic basis (Page and Fondrk 1996). Our behavioral data on selected lines also suggest that pollen and nectar collection are genetically correlated traits; selection on pollen foraging levels had a clear effect on nectar collection. Nectar and pollen foraging show an inheritance mode similar to a classic Mendelian model, where the phenotypic expression of pollen collection is recessive to that of nectar collection.

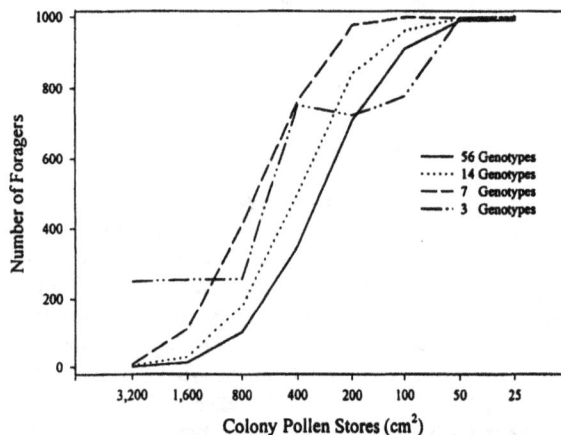

Figure 5. The expected change in colony pollen foraging levels based on a simulation model where 1000 workers are assigned intrinsic thresholds for pollen collection and subjected to increasing levels of stimulus for that task. The model varied from 3 to 56 genotypes for pollen thresholds, and allowed for phenotypic variation around those assigned genotypes. The colony mean threshold was set at 400 cm² of pollen, based on empirical studies. Pollen levels were varied from 25 cm² (highest stimulus) to 3200 cm² (lowest stimulus: two full frames of pollen). We performed 1000 iterations per pollen level, providing a high probability that any worker will encounter the stimulus. The lines show the expected shape of the colony's response to increased need for pollen as affected by variation in the number of loci affecting the trait.

Fewell: Foraging task organization by honey bees

The principal evidence for this comes from the behavior of hybrid workers, who showed dominance for nectar collection, even when the stimulus for pollen collection was high. This is consistent with the work of Hunt et al. (1995) who found that a high proportion of the variation in pollen foraging behavior can be mapped genetically to two quantitative loci.

Behavioral genetic models generally assume that phenotypic expression of a trait is continuous and based on additive genetic variation. Robinson and Page (1989a) use additive genetic variation as a starting point for their discussion of the stimulus threshold model, and Bonabeau et al. (1996) assume continuous variation in the probability of task performance in their mathematical consideration of the threshold model in ant task regulation. Given the variance between this and our current understanding of the genetic basis to foraging task preference, it is useful to evaluate how underlying genetic assumptions may affect the model.

Integration of genetic and behavioral models of task regulation

Our data offer strong support for the central tenet of the stimulus threshold model, that there is a direct relationship between genotype and task preference, and that genotypic diversity influences foraging flexibility. However, it is unlikely that a genetic model alone is sufficient to explain foraging regulation. Two factors affecting colony flexibility are individual flexibility in effort and mechanisms of recruitment to a given task. Because the threshold model focuses on allocation rather than individual effort, I will not discuss variation in individual foraging effort in this chapter, but it should not be ignored as a contributing factor to colony flexibility. In previous research in which we manipulated pollen stores in free-foraging colonies, Mark Winston and I found that a majority of colony response can be attributed to flexibility in individual pollen foraging effort, including foraging rate and load size (Fewell and Winston 1992). However, behavioral mechanisms of recruitment can also critically affect colony flexibility via worker task allocation, and the potential integration between these models and genotypic mechanisms of task regulation should be considered.

There are a variety of behavioral task regulation models, but we can divide them simplistically into

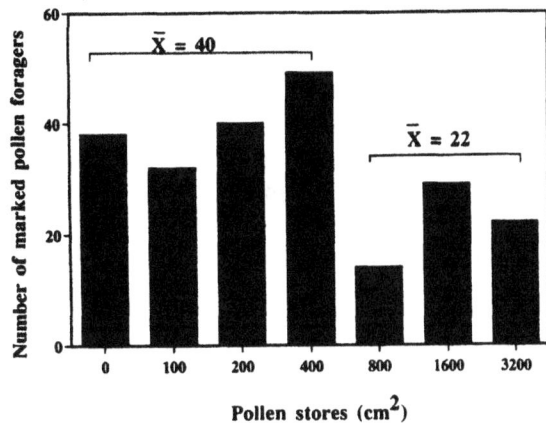

Figure 6. Number of marked workers collecting pollen as the amount of stored pollen stored was varied in two colonies. Pollen levels were gradually increased in the first colony and decreased in the second colony in a series of 7 manipulations performed every two days. Data are pooled for the two colonies, because the shape of colony response was similar. Both colonies showed a stepwise change in behavior as storage levels moved between 400 and 800 cm² of pollen.

two categories based on their expectations of how individuals receive information about tasks. In one category variation in task performance occurs primarily through variation in worker interactions with the hive environment. Workers perform different tasks because they encounter different task cues. Although this is an extremely simple assumption, random task encounter models can generate extremely complex patterns at the colony level. For example, Tofts and Franks (1992; Franks and Tofts 1994) argue that a simple system of variation in task encounter rates across a nest can generate age-related variation in task encounter and performance (age polyethism).

A second set of models make the expectation that task recruitment is based primarily on worker communication of task need or opportunity. Information transfer about task need or availability does not require a physical information "processing center", but instead occurs as a result of social interactions between workers engaged in the task and workers available to perform it (Seeley *et al.* 1991). In contrast to a random encounter model, information transfer through worker interactions allows the colony to respond in a coordinated and rapid manner to small changes in task stimuli, because information is actively transferred among workers

(Seeley 1986, 1996; Seeley and Levien 1987; Seeley *et al.* 1991).

The different behavioral models of task regulation vary in their predictions for how colonies respond to changes in task need. Because of this, we can tease them apart by gradually varying colony need for a specific task, such as pollen foraging and measuring the pattern of colony response. Specifically, we can ask the question of whether colonies respond in a graded manner to gradual changes in pollen need, or whether they show rapid or stepwise responses to relatively small gradations?

What are the predictions of colony behavior if the stimulus threshold model is combined with a model of random task encounter? If there are a large number of genotypes, and phenotypes are continuously distributed, then the response to a gradual increase in task need will be a gradual increase in the number of workers performing the task as an increasing number of thresholds are met. If the genotypes are few and discontinuous, the number of workers performing the task may still gradually increase, because, as long as there are more workers than task opportunities, worker task encounter rate will increase in direct relation to the number of opportunities to perform it.

If the stimulus threshold model is combined with an information transfer model and if worker task preference is continuously distributed, we should still see a graded increase in the number of workers performing the task. The pattern of response is constrained by the level of stimulus rather than how rapidly workers receive it, because workers only perform the task when their thresholds are met. However, if worker variation in task sensitivity is more bimodally distributed into specialist and nonspecialist subsets, then response may be strongly influenced by how rapidly workers receive information about the task. If information on task need is based on social interactions, workers can receive information on changes in need almost simultaneously. This can produce a sharply graded or stepwise change in the number of workers performing the task, as allocation of workers quickly shifts between high and low levels as the need for a task exceeds or drops below some set point. An integrated model of bimodal variation and information transfer does not require that colonies respond in a stepwise manner, but it is the only one that specifically allows for such a rapid and coordinated response.

Modeling colony behavioral responses in a stimulus threshold context

The predictions of colony response pattern can be modeled using a more formal approach. With Sue Bertram at ASU, I generated a simple simulation model of colony response based on variation in the phenotypic distribution of task thresholds. In particular, we wanted to address the question of whether we actually should expect a change from a threshold to a graded response curve as we move from a more discrete or bimodal pattern of stimulus thresholds.

In this model we assigned 1000 bees stimulus thresholds, with a mean threshold of 400 cm2 of pollen stores in the colony. This number was based on the set-point for changes in colony pollen collection (see experiment 3 below). We then randomly chose bees, over 1000 iterations, and exposed them to a given stimulus level. The high number of iterations generated a high probability that a given bee would be exposed to the task stimulus; this is consistent with an information transfer model. We then varied the number of genotypic subgroups for pollen foraging from three to 56, allowing for phenotypic variation around the mean response thresholds. As we moved from 3 to 56 genotypes, the variation in response threshold in our group of potential foragers became more continuous in distribution. As predicted, the shape of colony response to an increase in need for pollen moved from a stepwise to a more graded response as the number of behavioral categories increased (Figure 5). Additionally, the genotypic distribution of foragers shifted from low to high diversity as need for pollen increased, consistent with the stimulus threshold model and with the results from experiment 1.

Experiment 3: How do colonies respond to graded changes in pollen need?

Our current understanding of nectar foraging seems to support the predicted effect of an integrated genetic/information transfer model on colony response. A system of information transfer for communication about nectar availability (the information center model) was proposed and tested by Seeley *et al.* (1991). Consistent with expectation, honey bee colonies switched between resources rapidly as the relative quality of those resources varies. Nectar foragers regulate foraging rates primarily around resource availability and quality (Seeley 1986, 1989; Fewell and Winston 1996), and

they receive information on nectar resources through social interactions on the dance floor. Therefore, our understanding of inheritance patterns for nectar preference (fairly bimodal) and the mechanisms for transfer of stimulus cues for nectar foraging (socially transferred) are consistent with the expectation of a stepwise response curve.

Unfortunately, our ability to predict which model offers the best fit for pollen foraging is hampered by our lack of information on the specific cues for pollen foraging. Environmental stimulus cues for pollen collection may come from encounter of empty cells in the brood area (Camazine 1991; 1993), or from encountering brood itself. Contexts in which information transfer may occur include the hive dance floor (although this may be a site of information about pollen availability rather than need), and interactions with nurse bees, who process the pollen and may transfer information on its availability (Camazine pers comm).

With Sue Bertram, I performed a third set of cage experiments in which we asked two questions. First, does the genotypic diversity of the pollen foraging group vary as predicted by the stimulus threshold model? Second, what is the shape of the colonies' response to a graded change in pollen storage levels? Our experimental protocol for testing this model was similar to the previous experiments, except that instead of giving colonies the endpoints of high or low stores, we gradually changed the amount of pollen in the colony. We did this by either doubling or halving pollen storage levels in each of two colonies every two days, moving storage levels between 0 and 3200 cm2 (2 full frames) in 7 gradations (Fewell and Bertram, submitted). The manipulations were performed in different directions to control for environmental or developmental effects on worker behavior. The two colonies showed almost identical responses. In each case there was actually a slight increase in activity levels until pollen storage levels reached 400 cm2, about a half of one side of a pollen frame. As pollen levels exceeded this point, foraging behavior dramatically decreased (Figure 6).

This response occurred in both colonies, although the genotypic mix of marked workers in the two colonies was completely different. The strong coordinated responses to changes in pollen need are consistent with an information transfer model. However, genotypic variation remained a contributing factor to task allocation. Only one of the two colonies showed significant variation among the three marked groups in foraging preference. As that colony moved from low to high pollen need, we saw the predicted shift from pollen foragers being over-represented by one group to a much more even distribution of worker genotypes.

Conclusions

The data presented here provide strong cumulative support for the hypothesis that genetic variation plays a central role in task allocation in honey bees. Models of task organization in social insects other than honey bees have largely ignored genetic variation, but recent work on ants (Stuart and Page 1991; Snyder 1993) and social wasps (O'Donnell 1996) demonstrates that these groups also show a strong link between genotype and individual task choice. Therefore it is likely that the genetic models being developed for honey bees will be broadly applicable across social groups.

These experiments strongly support the tenet of the stimulus threshold model that genotypic diversity plays a central role in colony task organization. However, we cannot completely explain colony task regulation through genetic variation alone. Division of labor in the context of foraging may best be explained by an integrated model incorporating both genetic and environmental components. In such a model, a subset of workers become genetic specialists for pollen collection; other foragers are likely to collect nectar, but can be recruited into pollen foraging as need increases. The colony is able to continuously monitor the resource environment through these specialists, allowing it to respond in a rapid and coordinated manner as need or availability changes. Such a model is probably still a somewhat simplistic picture of colony function, but it fits with our emerging understanding of how simple rules of behavior can generate the coordinated and flexible design so beautifully illustrated by the honey bee colony.

Acknowledgments

I thank Jon Harrison, Rob Page, Sue Bertram, Glennis Julian and the many students of my lab for their help on these projects and for numerous discussions of this topic. They may not always have agreed with me, but they certainly encouraged me to think! This work was supported by NIMH Grant R29-MH51329.

References

Bonabeau, E., G. Theraulaz, and J.L. Deneubourg. 1996. Quantitative study of the fixed threshold model for the regulation of division of labor in insect societies. *Proc. R. Soc. Lond. B* 263:1565-1569.

Calderone, N.W., and R.E. Page Jr. 1988. Genotypic variability in age polyethism and task specialization in the honey bee, *Apis mellifera* (Hymenoptera: Apidae). *Behav. Ecol.Sociobiol.* 22:17-25

Calderone, N.W., and R.E. Page Jr. 1991. Evolutionary genetics of division of labor in colonies of the honey bee (*Apis mellifera*). *Amer. Nat.* 138:69-92.

Camazine, S. 1991. Pattern formation on the combs of honey bee colonies: self-organization based on simple behavioral rules. *Behav. Ecol. Sociobiol.* 28:61.

Camzine, S. 1993. The regulation of pollen foraging by honey bees: how foragers assess the colony's need for pollen. *Behav. Ecol. Sociobiol.* 32:265-272.

Eckert, C.D., M.L. Winston and R.C. Ydenberg. 1994. The relationship between population size, amount of brood, and individual foraging behaviour in the honey bee, *Apis mellifera* L. *Oecologia* 97:248-255.

Fewell, J.H. and S.M. Bertram. Division of labor in a dynamic environment: response to graded changes in colony pollen stores by honey bees. Submitted manuscript.

Fewell, J.H. and R.E. Page Jr. 1993. Genotypic variation in resource selection by honey bees, *Apis mellifera. Experientia* 49:1106-1112.

Fewell, J.H. and R.E. Page Jr. Colony-level selection effects on individual and colony foraging task performance in honey bees, *Apis mellifera* L. submitted manuscript.

Fewell, J.H. and M.L. Winston. 1992. Colony state and regulation of pollen foraging in the honey bee, *Apis mellifera* L. *Behav. Ecol. Sociobiol.* 30:387-393.

Fewell, J.H. and M.L. Winston. 1996. Regulation of nectar collection in relation to honey storage levels by honey bees, *Apis mellifera. Behav. Ecol.* 7:286-291.

Franks, N.R. and C. Tofts. 1994. Foraging for work: how tasks allocate workers. *Anim. Behav.* 48:470-472.

Free, J.B. 1967. Factors determining the collection of pollen by honeybee foragers. *Anim. Behav.* 15:134-144.

Frumhoff, P.C. and J. Baker J. 1988. A genetic component to division of labour within honey bee colonies. *Nature* 333:358-361.

Hunt, G., R.E. Page Jr, M.K. Fondrk and C.J. Dullum. 1995. Major quantitative trait loci affecting honey bee foraging behavior. *Genetics* 141:1537-1545.

O'Donnell, S. 1996. RAPD markers suggest genotypic effects on forager specialization in a eusocial wasp. *Behav. Ecol. Sociobiol.* 38:83-88.

Oldroyd, B.P., T.E. Rinderer and S.M. Buco. 1992. Intracolonial foraging specialism by honey bees (*Apis mellifera*; Hymenoptera: Apidae). *Behav. Ecol. Sociobiol.* 30:291-295.

Page, R.E. Jr. 1980 The evolution of multiple mating behavior by honey bee queens (*Apis mellifera* L.). *Genetics* 96:263-273.

Page, R.E. Jr and M.K. Fondrk. 1996. The effects of colony-level selection on the social organization of honey bee (*Apis mellifera* L.) colonies: colony-level components of pollen hoarding. *Behav. Ecol. Sociobiol.* 36:135-144.

Page, R.E. Jr and G.E. Robinson. 1991. The genetics of division of labour in honey bee colonies. *Adv. Insect Physiol.* 23:117-171.

Robinson, G.E. and R.E. Page Jr. 1988. Genetic determination of guarding and undertaking in honeybee colonies. *Nature* 333:356-358.

Robinson, G.E. and R.E. Page Jr. 1989a. Genetic basis for division of labor in an insect society. *In:* M.D. Breed and R.E. Page, eds. The Genetics of Social Evolution. Westview Press, Boulder Colorado, pp 61-80.

Robinson, G.E. and R.E. Page Jr. 1989b. Genetic determination of nectar foraging, pollen foraging, and nest-site scouting in honey bee colonies. *Behav. Ecol. Sociobiol.* 24:317-323.

Seeley, T.D. 1986. Social foraging by honeybees: how colonies allocate foragers among patches of flowers. *Behav. Ecol. Sociobiol.* 19:343-354.

Seeley, T.D. 1989. Social foraging by bees: how nectar foragers assess their colony's nutritional status. *Behav. Ecol. Sociobiol.* 24:181-199.

Seeley, T.D. and R.A. Levien. 1987. Social foraging by honeybees: how a colony tracks rich sources of nectar. In: R.Menzel and A. Mercer, eds. Neurobiology and Behavior of Honeybees. Springer Verlag, pp38-53.

Seeley, T.D., S. Camazine and J. Sneyd. 1991. Collective decision-making in honey bees: how colonies choose among nectar sources. *Behav. Ecol. Sociobiol.* 28:277-290

Snyder, L. 1993. Non-random behavioural interactions among genetic subgroups in a polygynous ant. *Anim. Behav.* 46:431-439

Stuart, R. and R.E. Page Jr. 1991. Genetic component to division of labor among workers of a Leptothoracine Ant. *Naturwissenschaften* 78:375-377.

Tofts, C. and N.R. Franks. 1992. Doing the right thing: ants, honeybees and naked mole-rats. *Trends Ecol. Evol.* 7:346-349.

Winston, M.L. 1987. The Biology of the Honey Bee. Harvard University Press, Cambridge, MA.

Regulation of division of labor in worker honey bees: the activator-inhibitor model

ZACHARY Y. HUANG
DEPARTMENT OF ENTOMOLOGY
MICHIGAN STATE UNIVERSITY
EAST LANSING MI 48824
Tel. 517-353-8136
Fax: 517-353-4354
E-mail: bees@msu.edu

Summary

Honey bee workers change jobs as they age. The age polyethism of workers has previously been shown to have a high degree of plasticity. For example, workers normally start their foraging career when they are around 3 weeks of age; however, when a colony has no foragers, workers can start foraging at only one week of age. Conversely, workers older than 3 weeks can remain as "overage" nurses when no young bees are emerging in a colony. What is the mechanism that regulates the onset of foraging in workers? Through rearing workers under different social conditions and manipulating the age structure of a colony, it is found that workers regulate their onset of foraging through a simple negative feed-back mechanism, which we called an "activator-inhibitor" model. The activator level in a worker determines the probability of it becoming a forager. Workers emerge with low levels of the activator, which is programmed to increase. Once the activator reaches a critical level in a worker, the worker becomes a forager. The inhibitor level also becomes high in foragers, so that other workers are inhibited from becoming foragers. This model is robust because three major predictions by the model are met with vigorous experimental testing. Workers became foragers faster when level of the presumed inhibitor was decreased in a colony, they became foragers slower when the level of inhibitor was increased. Workers with high juvenile hormone titers decreased their hormonal levels when the level of inhibitor was drastically increased. The activator-inhibitor model provides a heuristic tool for understanding the division of labor in honey bee colonies.

I. Introduction

1. Division of labor

Division of labor has been thought as a major factor for the enormous success of social insects (Wilson 1985). Presumably, division of labor increases the efficiency of an organization because it allows many different tasks to be performed simultaneously by many different individuals (Oster and Wilson 1978). The most common form of division of labor in social insects is reproductive division of labor: the queen reproduces almost exclusively, while other non-reproducing members specialize in the day to day operation and maintenance of the colony. In many insect societies, there is a further division of labor among the non-reproducing members. This can be based on several mechanisms. In some species of termites and ants, some members are morphologically specialized to defend the colony and are called "soldiers," while the others are called workers. Workers in some ant species show large size variations and specialize in different tasks (Hölldobler and Wilson 1990). In many advanced social insects, division of labor is related to age: workers of different ages specialize in different tasks. This phenomenon is called "age polyethism."

2. Age polyethism in honey bees

In honey bees, workers typically perform brood rearing ("nursing") for the first week, engage in other hive maintenance duties when they are "middle-aged" (2-3 weeks old), and switch to foraging and colony defense when they are about three weeks old. Despite this typical pattern of behavioral development, age polyethism is also highly flexible (reviewed by Robinson 1992). Individuals can accelerate, delay, or even reverse their behavioral development in response to changes in the colony's internal and external environment. Workers can start foraging as early as one week old, if a colony is artificially made with only newly emerged bees (Nelson 1927, Milojévic 1940, Robinson *et al.* 1989). These workers are called precocious foragers because of their young age. Precocious foragers can also be observed in natural colonies (Ribbands 1952), although only infrequently. Conversely, workers can become foragers when they are much older than 3 weeks. One extreme example is bees in temperate areas (in the mid-western United States, for example): workers that emerge during October never have a chance to become foragers in the fall. They live for at least five months, because the earliest flowers would only be available in March or April the next year. Delayed development also occurs in a new colony founded by a swarm, because new workers would not emerge until at least 21 days after the queen starts laying eggs, so the youngest nurses around that time would be at least 21 days old. Even more striking is "reversion", when colony age demography is drastically changed. In this case, old workers become physiologically "younger" and engage in behaviors that are normally performed by younger bees. For example, one can remove all the young bees in a colony, and leave only foragers. In this colony, bees that have been observed foraging on the previous day revert back to taking care of the queen and brood (Milojévic 1940, Page *et al.* 1992).

3. Effect of Juvenile hormone on age polyethism

Juvenile hormone (JH) is a major developmental hormone involved in the regulation of insect metamorphosis, the process that changes a caterpillar into a butterfly. In an insect larva, high JH titer in blood causes the next developmental stage to remain larval (rather than pupal or adult), therefore it appears to keep the immature insect "juvenile." The name juvenile hormone often brings confusion

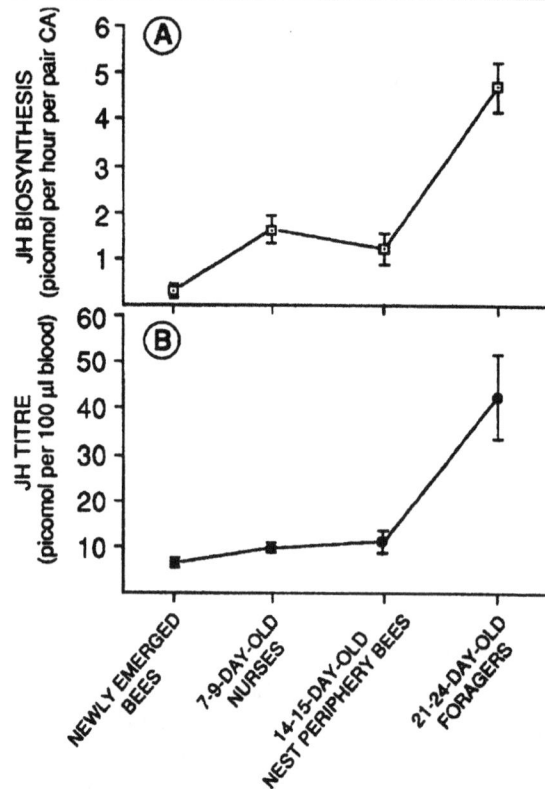

Figure 1. Changes of juvenile hormone biosynthesis (A) and titer (B) with age in worker honey bees, presented as mean ± standard error. Rates of JH biosynthesis determined in individual bees (n=10), titer determined as pooled blood samples (5-8 bees per sample) (n = 5-8 samples), using the Strambi RIA method (Strambi et al. 1984). Now it is possible to measure JH titer in individual workers, with as little as 0.5 μl of blood (Huang and Robinson 1995, 1996). (Reprinted from *Journal of Insect Physiology*, Vol. 37, ©1991, Page 733-741, with permission from Elsevier Science).

because the effect of JH in adult worker bees is quite the contrary.

JH possesses a maturing effect in the behavioral development of adult honey bee workers, instead of the juvenilizing effect found in metamorphosis (reviewed by Fahrbach and Robinson 1996, Robinson and Vargo 1997). JH titers in blood typically increase with age; they are low in bees that perform in-hive tasks such as nursing, comb building and other activities, and high in foragers (*Figure* 1) (Rutz *et al.* 1976; Fluri *et al.* 1982; Robinson *et al.* 1987, 1989; Huang *et al.* 1991, 1994; Huang and Robinson 1995). Figure 1 also shows that, when corpora allata, the organs that produce JH, are removed from workers and incubated in vitro (outside the body in test tubes), they produce amounts of JH that are correlated with blood JH titers in intact bees. Rates of JH biosynthesis show the same upward change with age, and are low in young bees, and high in foragers. The methods to measure JH titers and rates of biosynthesis are presented in detail in the Appendix.

However, a simple association of one chemical with an event, be that event a behavioral or physiological one, is not enough to conclude that the chemical causes the event. If JH causes the shift in-hive tasks to foraging, then artificially elevating the JH titers should cause workers to become foragers earlier. Indeed, applying JH or JH analog (chemicals that have similar structure and function as JH) to workers causes them to forage earlier (Jaycox 1976; Jaycox et al. 1974; Robinson 1985, 1987; Robinson and Ratnieks 1987; Robinson et al. 1989; Sasagawa et al. 1989).

Not only is JH involved in the regulation of the normal behavioral development of workers, it also is implicated in the flexibility of behavioral development mentioned earlier. When one-week-old bees become foragers due to a lack of old bees in the colony, these precocious foragers also show high JH titers, despite their young age. When behavioral development is delayed, the overage nurses have low titers, despite their old age. Finally, bees that revert from foraging to nursing show a drop in JH titer (Robinson et al. 1989; 1992, Huang and Robinson 1996), despite the fact that JH usually rises in workers. These results support the idea that changes in colony conditions act on the JH levels to cause changes in behavioral development (Robinson 1987).

4. Mechanism of regulation of behavioral development

While much is known about JH and its relationship with the onset of foraging, it is not clear how workers "know" when to start foraging. Many studies on the age polyethism of bees have focused on the onset of foraging because this behavioral change is the most dramatic and easy to observe. Differences in the onset of foraging accurately reflect overall differences in the behavioral development because foraging happens to be the final task performed by workers. Over the years, many different factors have been suggested to affect the time of foraging onset. Age of first foraging has been proposed to be affected by worker loss, amount of open brood, and colony size. These factors, however, seem not to affect age of first foraging consistently. For instance, it was first discovered that bees in smaller colonies foraged earlier (Winston and Punnett 1982, Winston and Fergusson 1985), but in another study foragers developed faster in larger colonies (Neumann and Winston 1990). Similarly, bees foraged earlier in colonies with less brood (Winston and Punnett 1982, Winston and Fergusson 1985), but no brood effect was found in later studies (Winston and Fergusson 1986, Neumann and Winston 1990). Fukuda (1960) suggested that pollen shortage could lead to accelerated behavioral development, but the early foraging could easily be caused by differences in genetic makeup and colony conditions, because only two colonies (one treatment colony and one control) were used.

In this review, I will discuss experiments that I performed in collaboration with Gene Robinson. These experiments point to social interaction as a crucial factor in regulating hormonal and behavioral development in workers, leading to the formulation of the activator-inhibitor model. I then present a description of the model, a detailed review of experiments to test the model, and the implication of the model.

II. Regulation of division of labor via social interaction

As shown in Figure 1, JH titers or rates of biosynthesis increase with age, such that it is high in foragers, and low in bees performing tasks inside the hive. However, this relationship is only found in typical colonies with a natural age demography. Workers can have high JH levels when they are seven days old if they become precocious foragers; conversely, three week old workers can have low JH levels if they are still nursing. Age itself, therefore, does not seem to be the only determinant for JH change in workers. To dissect the mechanism for regulating the hormonal and behavioral development in worker bees, we conceptually divided the colony environment into two parts: nest environment and social environment. Workers in a colony composed of only new emerged bees (a single cohort colony) could either sense the lack of fresh incoming nectar and pollen, normally present when there are foragers around, or directly sense the absence of foragers, and then "decide" to become precocious foragers.

1. Isolation experiment

In this experiment we examined whether the nest environment or the social environment plays a more important role in regulating the JH changes. Newly emerged workers were reared under three different conditions. One group of bees was reared inside a typical colony, so that they had access to both the social and nest environment. A second group was reared in small groups in the laboratory, where they had a type of social environment; because any time two bees are together, they interact with each other behaviorally (e.g., mutual grooming and feeding). Bees in a third group were reared as isolated individuals and provided with the same food (sugar candy) as those reared in a group. Bees in the third group had access to neither the social nor the nest environment. After one week, the isolated workers showed prematurely elevated rates of JH biosynthesis, similar to those of normally aged foragers (Figure 2). In contrast, bees reared in groups showed similar rates of JH biosynthesis to bees in a normal colony. The high JH levels in isolated bees were not due to an absence of nest environment, but rather an absence of social environment. This is because bees reared in small groups showed rates of JH biosynthesis typical of 7-day-old colony-reared bees, even though they lacked exposure to any nest stimuli.

There is also a "group-size" effect on the rates of JH biosynthesis. As the number of workers per group increased, the rates of JH biosynthesis in each group decreased drastically (Huang and Robinson 1992). Figure 3 shows that the percentage of workers showing "forager-like" rates of JH biosynthesis (2.84 picomole per hour per pair of corpora allata) decreased dramatically as the number of workers in each group increased. Assuming that more bees would give rise to more bouts of social interactions, this result suggests that social interaction also has a quantitative effect on JH levels.

One could perhaps argue that the isolated bees were stressed because the isolation does not occur naturally. To see if their behaviors are consistent with their hormonal levels, we also compared their behavior with bees reared in a colony. When both groups of bees are introduced to a colony that contained other 7-day-old bees as "background" bees,

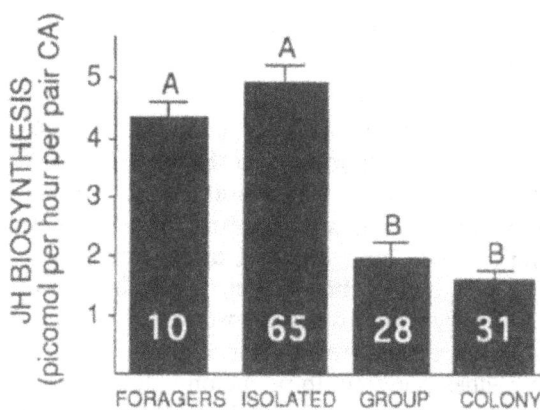

Figure 2. **Effect of social and nest environment on juvenile hormone biosynthesis (mean ± Standard error) in worker bees. Newly emerged workers were reared either in isolation ("isolated"), in groups of 4 bees ("group"), or in a colony ("colony"). Rates of juvenile hormone biosynthesis in vitro were measured after 7 days. Hormonal levels of normal aged foragers were also presented for comparison. Number on each bar indicates number of bees tested. Bars with different letters on top indicate a significant difference by Tukey's test at the 5% level (data from Huang and Robinson 1992).**

Huang: Regulation of division of labor

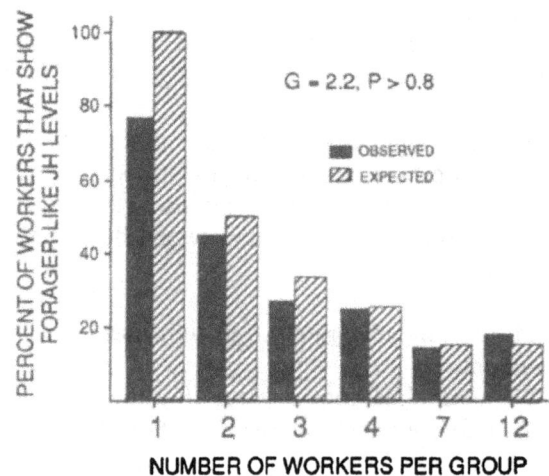

Figure 3. **Observed and expected percentage of workers that show forage-like JH levels in bees reared in different group sizes. Expected percentage was calculated as follows: one worker is assumed to be able to inhibit up to 6 other bees, but cannot inhibit itself: so the percentages of bees expected to behave hormonally as foragers for group size of 1, 2, 3, 4, 7, 12 are 100%, 50%, 33.33%, 25%, 14.3% and 14.3%, respectively (data from Huang and Robinson 1992).**

67% of bees reared in isolation became foragers, while only 1.2% of colony reared bees did so (Huang and Robinson 1992). These results suggest, but do not prove, that socially isolated bees are not abnormal due to the artificial rearing condition, but rather are behaving as foragers both hormonally and behaviorally, when they are only 7 days old.

2. "Transplant" experiment

The importance of nest vs. social environment was further tested in a colony setting by using a "transplant" assay. In this assay, we utilized the precocious development in single cohort colonies. We transplanted a group of foreign bees into a single-cohort colony and observed the age of first foraging in the resident bees. When foragers were used as the transplant, precocious development was inhibited in the resident bees, apparently due to the inhibition from the transplanted old bees (Figure 4A). This was a specific effect of transplanting foragers, because resident bees developed precociously when young bees were transplanted. Because the transplanted older bees were foragers, there was a possibility that their nectar or pollen loads could have been used by the resident bees as cues for the presence of old bees. This possibility was eliminated by transplanting "soldiers", bees that specialize in defending the colony from large animals. Soldiers have been shown to be behaviorally, genetically, and physiologically distinct from foragers (Breed *et al.* 1990, Huang *et al.* 1994). Soldiers inhibited behavioral development in a manner similar to foragers.

Was the inhibitory effect of foragers mediated by their interaction with resident bees, or by changes in the nest environment caused by their foraging activity? To test the two alternatives, we performed transplant experiments with the colony entrance closed (the colony was closed for 3 days and opened when observation for foraging started), thereby eliminating any change of the nest due to foraging. Inhibition of behavioral development in resident bees was still observed (Figure 4B). These results suggest that behavioral development of young bees is affected directly by the presence of foragers.

III. The activator-inhibitor model

Based on the results of the isolation and transplant experiments, a descriptive model was proposed to explain how social interactions influence behavioral development in honey bee workers

(Huang and Robinson 1992). An "activator" is assumed to increase the probability of becoming a forager. Obviously JH fits well in this role. There is also an "inhibitor" that is transferred among workers socially and perceived by workers, which subsequently hinders their increase of JH. The activator is assumed to be programmed to increase to the level of foragers in about a week, but in a normal colony this increase is delayed by the presence of inhibitors. The activator and inhibitor levels are coupled, such that when a bee has a high level of activator, it also is more capable of inhibiting other workers. Workers are not able to "self-inhibit" when they are alone because the inhibitor has to be transferred socially. According to this model, workers reared in isolation become forager-like in their hormone levels because no inhibitor is present; workers reared in a group have low JH levels because

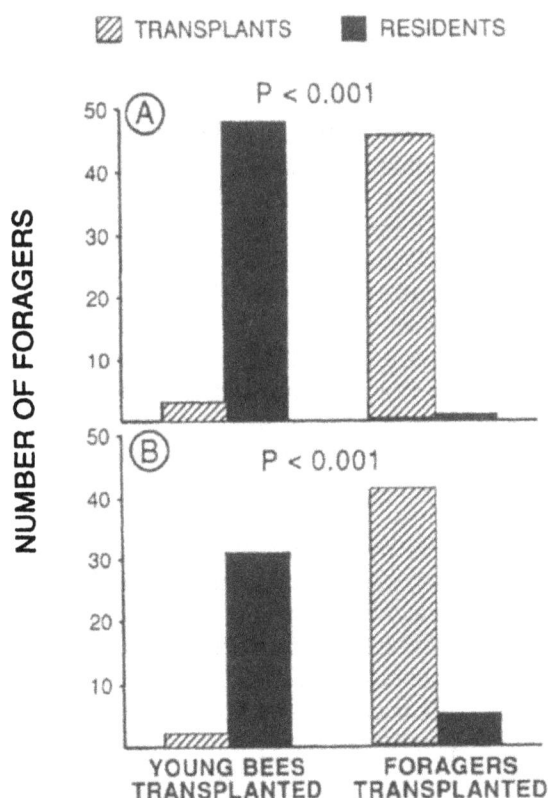

Figure 4. Effect of transplanted bees on number of resident bees that behave as foragers. A: Transplant composed of 20% pollen foragers and 80% soldiers, entrance open. B: Transplant composed of 100% soldiers, entrance closed for 4 days until observation for foraging began on day 8 (data from Huang and Robinson 1992).

only one or a few bees in each group have high JH levels and the rest are inhibited by these bees. The model predicts that as the number of bees in a group increases, the proportion of bees with "forager-like" rates of JH biosynthesis in each group would decrease, which is consistent with the results of the isolation experiment (Figure 3). Similarly, in a single cohort colony, some workers become precocious foragers because no inhibitor is present initially; but as soon as precocious foragers are produced, they possess high levels of inhibitor and prevent other workers from becoming foragers. Therefore one only observes 5-10% of bees becoming precocious foragers, not 100%, because the negative feed back system works as soon as precocious foragers are developed. In a colony with predominantly old bees, workers receive high levels of inhibitor from older workers, delaying their increase of JH, resulting in delayed development. According to this model, genetic variation for production of the factors (activator and inhibitor), and the sensitivity to them, can explain previously observed genetic variation in rates of behavioral development (Robinson *et al.* 1989, Kolmes *et al.* 1989, Giray and Robinson 1994).

The activator-inhibitor model explains nicely the precocious development of bees both in social isolation and a single cohort colony, and the results of transplant experiments. However, it is always more convincing and satisfying if a model can provide reliable predictions in more natural colony settings. In the following section, I discuss two experiments aimed at testing the model under more realistic colony conditions, and a third experiment that tests the model critically.

IV. Testing of the activator-inhibitor model

In the first two experiments, we used designer-colonies that allowed us to precisely control the genetic makeup, age structure, and food resources within each colonies. These colonies are called "triple-cohort colonies", each composed of approximately equal numbers of "young", "middle-age", and "old" bees. Young bees (newly emerged within the last 24 hours) were obtained by placing combs of sealed brood in an incubator at 34°C and 80% relative humidity. They were marked with a spot of paint (Testor's PLA) on the dorsal surface of the thorax. Middle-age bees were marked a different color and reared in a nursery colony and recovered around 10 to 12 days of age. Old bees were foragers, col-

lected by blocking the entrance of a typical colony and vacuuming bees that returned with pollen or nectar. These bees were of unknown ages, but foragers usually are the oldest bees in typical colonies (reviewed by Michener 1974; Winston 1987). Foragers were collected from a colony located ~10 km away from the triple-cohort colony to minimize their return to the natal colony. Middle-age bees were the "focal bees" (bees on which the experimental outcome was measured) for age of first foraging in experiments 1 and 2.

The triple-cohort colonies we used were composed only of 1,500 to 2,000 bees, which are small compared with typical natural colony sizes (15,000-40,000, Seeley 1985). Previous experiments using the same technique, however, showed that workers in these colonies exhibit normal age polyethism (Giray and Robinson 1994).

1. Effects of decreasing inhibitor level on behavioral development

In the first experiment, two colonies of similar genetic makeup and age demography were established. In one colony, returning foragers were removed to simulate predation, while the control colony had the same number of bees removed, but across all three age cohorts. The two colonies therefore had the same population size, but different age structure. According to the activator-inhibitor model, young bees in the forager-depleted colony should start foraging earlier, because the amount of inhibitor in this colony is reduced due to forager removal. Indeed, in all three replicates of this experiment, more middle-aged bees became foragers in the forager-depleted colony than the control colony (Figure 5A). Decreasing the inhibitor level in a colony caused workers to become foragers faster, as predicted by the activator-inhibitor model.

2. Effects of increasing inhibitor level on behavioral development

In a second experiment, we confined foragers most of the day (5:30 am to 8 pm) inside a colony for a few days (3, 4, and 9 days for trials 1, 2, and 3, respectively) using a water sprinkler. After this period, the colony was opened and number of foragers from the focal group was quantified. A control colony was setup at the same time but normal foraging was allowed. We predicted that workers in this colony would delay their development because of two factors. First, when foragers were confined, they should transfer more inhibitor to other bees, because they spent 100% percent of their time inside the colony, while foragers in the control colony spent portions of their time foraging in the field. Second, because mortality of foragers was decreased (most foragers die during foraging trips due to predation or attrition), the confined colony should have a higher proportion of foragers than the control colony at the end of the experiment.

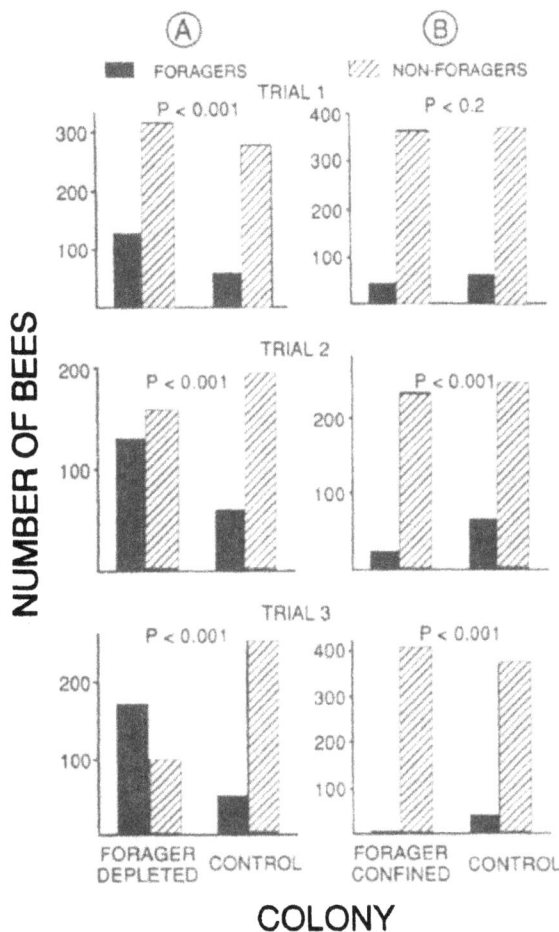

Figure 5. Number of bees from the middle-age cohort in triple cohort colonies that did or did not initiate foraging in response to alterations in colony age demography. A: Experiment 1 - forager depletion. Equal numbers of bees were taken from control colonies, but distributed equally across all three age cohorts. B: Experiment 2 - forager confinement with simulated rain. Control colonies were unmanipulated (Reprinted from *Behavioral Ecology and Sociobiology*, Vol. 39, ©1996, Page 147-158, with permission from Springer-Verlag).

We found that in two out of three replicates of this experiment, workers decreased their rate of behavioral development. A smaller proportion of focal bees became foragers in the forager-confined colony than the control colony during the same period (Figure 5B). This result is consistent with the prediction of the activator-inhibitor model. The results of the first trial (i.e. no difference between the confined and control colony) also lends anecdotal support for the model. During the three days of water sprinkling of the forager-confined colony, foragers in the control colony were also confined by cold and rainy weather; so in essence the control colony became "treated" inadvertently due to environmental conditions.

The results of the first experiment also are consistent with a competing hypothesis: that depleting foragers also reduced the amount of incoming resources such as nectar or pollen, so workers perhaps responded to these changes by foraging earlier. This "resource hypothesis" predicts that workers should also speed up in the forager-confined colony, perhaps even more drastically than what was observed in Experiment 1, given that no fresh food sources were brought into this colony. In the second experiment, the alternative hypothesis is not supported because we observed a delayed, rather than an accelerated, behavioral development in workers.

3. Effects of drastically increasing inhibitor level on hormonal development

As mentioned previously, some foragers show a decrease of JH titers in a colony made up entirely of foragers (Page *et al.* 1992, Robinson *et al.* 1992). However, because open brood was in the colonies in these studies, it was thought that the observed reversion is due to stimuli from brood. It is reasonable to assume that in a colony with no nurses, larvae would become starved and foragers would receive enhanced stimuli for feeding them (see Huang and Otis 1991), causing foragers to revert to nursing. After the activator-inhibitor model was developed, we realized that according to the model, hormonal reversion could be solely caused by the super-abundance of the inhibitor—because every worker could interact only with other foragers, all of whom have high levels of inhibitor. Foragers would be "overdosed" with inhibitor and their hormonal levels should decrease. An experiment to observe hormonal decrease in a colony without brood would be a litmus test for the model, because this is an entirely new interpretation for an already observed phenomenon. But, how could we identify the "reverted" bees if no brood was in the hive? To accomplish this we introduced a frame of open brood to the colony for 10 minutes and sampled bees that were visiting larvae, after which the brood frame was removed. It is possible that by mistake a few young bees were collected into these "all-forager" colonies. These bees would have low JH titers and also a higher tendency to put their heads in the larval cells. This would create an illusion of "reversion" while there was none. To eliminate this possibility, a frame of open brood was used to collect workers immediately after the reversion colony was made. If there was young-bee contamination, one should see low JH titers in these first samples. As shown in Figure 6A, bees sampled 10 minutes after the colony establishment showed similar hormonal levels as returning foragers; however 24 hours later bees sampled using the same technique showed a much lower JH titer (Fig 6A inset). In a second colony, we waited 2.5 hours after the colony was made to collect the first sample. Interestingly, the JH of these bees were already significantly lower than that of the foragers (Figure 6B). This suggests that hormonal reversion occurs relatively fast, although further replication is needed to make a firm conclusion. To guard against the possibility that the 10 minute brood presentation might cause the observed reversion after 24 hours (if the signal from brood was very powerful), we did not present a brood frame to take the first sample in the third colony. Hormonal titer still drastically dropped after 24 hours in this colony (Figure 6C inset), with no exposure of workers to brood. Taken together these results suggest that indeed hormonal reversion can be induced simply by manipulation of colony age demography in the absence of brood, which again provides strong support for the model.

One unexpected outcome of this third experiment is the change of another physiological parameter associated with the change of JH titers. We also measured the size of hypopharyngeal glands in bees whose blood was sampled for JH measurement. Hypopharyngeal glands are glands located in the head of workers. They are relevant to division of labor because the glands are well developed and secrete "royal jelly" proteins in nurse bees; while in foragers they switch to the production of invertase and become regressed in size (Jung-Hoffman 1966,

Simpson *et al.* 1968). Reverted bees regenerated their small hypopharyngeal glands in all three colonies, while bees that continued foraging did not (Figure 7). The significance of this finding is discussed in the section below.

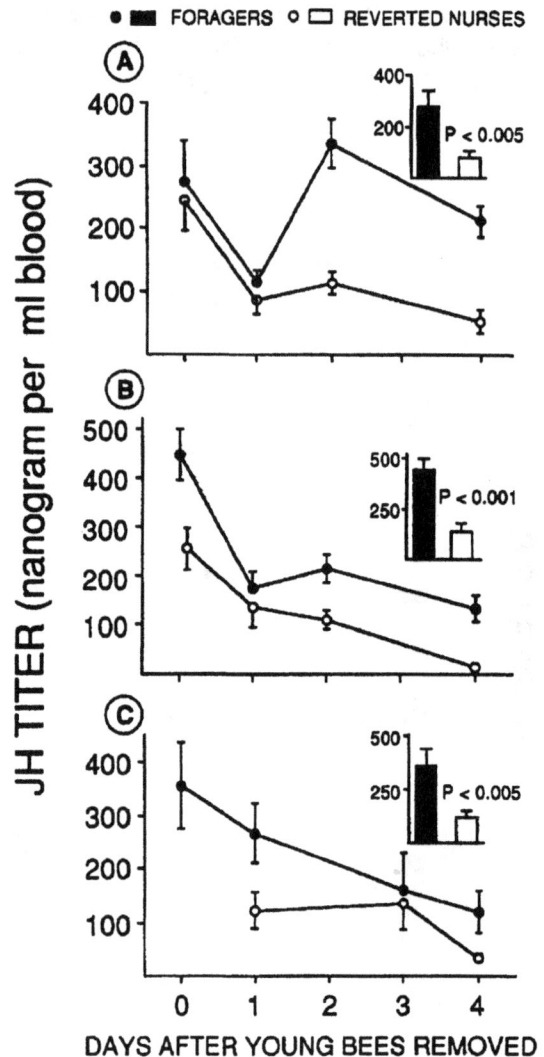

Figure 6. Changes in mean (±SE) blood titer of JH for foragers and reverted nurses in response to removal of young bees in the absence of brood (N = 10 for each point). Reverted nurses on day 0 were collected 10 min and 2.5 h after young bees were removed, in trials 1 and 2, respectively. A comparably early sample was not taken in trial 3. Inset: differences in JH titer between foragers on day 0, just before young bees were removed, and reverted nurses on day 1. P values from one-tailed t-tests (Reprinted from *Behavioral Ecology and Sociobiology*, Vol. 39, ©1996, Page 147-158, with permission from Springer-Verlag).

V. Implications of the activator-inhibitor model

1. Maintaining the colony structure via the activator-inhibitor model

Recently there has been a controversy over whether age-related polyethism in social insects is simply an epiphenomenon, or a developmental process. The "foraging-for-work" model asserts that workers failing to find a job at a certain nest location simply move on to the next task that is not crowded with workers (Tofts and Franks 1992, Franks and Tofts 1994). This creates a correlation between age and tasks performed, but purely due to the mechanics of young bees continually emerging and displacing workers to the next task, until they eventually begin foraging. The opposing side argues that there are physiological, biochemical and neuroanatomical changes associated with age of workers, and these changes affect workers' probability of task performance (Robinson *et al.* 1994). A critical test was proposed to settle this controversy, which is to have "a demonstration that physiological changes occur prior to a change in the worker's task, when task demand was held constant" (Franks and Tofts 1994). The third experiment (hormonal reversion without brood) discussed above meets this requirement. The task demand for brood rearing was held constant (at zero), no brood was reared in these colonies because the queen was caged and brood was removed. Despite the lack of brood, physiological changes (lowering of JH titers and the enlargement of hypopharyngeal glands) still occurred to make the foragers more suitable for nursing duties.

These results suggest that workers maintain a social structure with different physiological status, even when tasks requiring these physiological status are absent. The dissociation of social structure from task demand has several advantages. First, maintaining a social structure may save energy compared to switching back and forth among different tasks. For example, switching from nursing to foraging and vice versa may be energetically costly, because workers need different exocrine gland status to perform these two different tasks. A second benefit is that it may allow a faster response when the task demand is eventually presented. There is no need for bees to change their physiological status, because they are already physiologically "ready". According to this view, a colony of bees should still allocate a certain portion of its work force as foragers, even when foraging is restricted due to weather or dearth of resources. In fact, bees have been observed to forage for water and pollen in the middle of winter (February 22, 1992), when it became unusually warm (Huang and Robinson 1995). Conversely, when there is no brood present in a colony, some members still maintain their physiological competency at nursing. If nurses are lacking, bees performing other tasks would "anticipate" the need for nursing and develop their physiological competency for that task, as was shown in the reversion experiment.

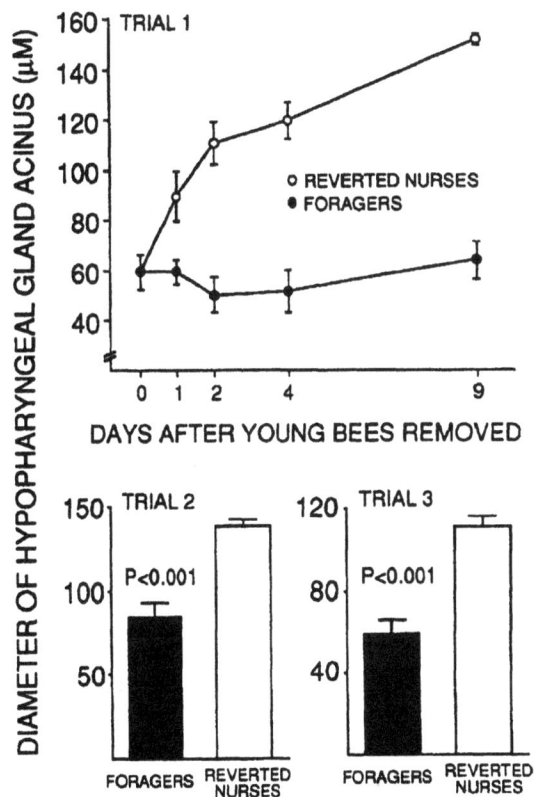

Figure 7. Changes in mean (±SE) hypopharyngeal gland size for foragers and reverted nurses in response to removal of young bees, in the absence of brood (N = 10 for each point). Bees were sampled 22 and 7 days after young bees were removed, in trials 2 and 3, respectively (Reprinted from *Behavioral Ecology and Sociobiology*, Vol. 39, ©1996, Page 147-158, with permission from Springer-Verlag).

2. The search for the inhibitor

The most obvious prediction of the activator-inhibitor model is that there should be an inhibitor which prevents young bees from growing up into foragers. Recent studies suggest that the inhibitor delaying behavioral development is produced in the mandibular glands, and this primer pheromone is transmitted via direct social contact (Huang et al, 1998). As shown in the transplant experiment, old workers (foragers or soldiers) can inhibit precocious foraging in a single cohort colony. When mandibular glands were removed surgically, the old bees showed significantly less inhibition. This reduction of inhibition appears due to gland removal, because sham-operated bees (bees that had incisions made on cuticles but glands were not removed) did not show the same reduction. In another set of experiments, newly emerged bees were reared for 7 days in a typical colony in one of three ways: individually in cages with a double screen, individually in cages with single screen, or freely inside the hive (control bees). Bees in double-screen cages experienced the same odor environment, but had no direct physical contact with other bees; bees in single-screen cages experienced the odor environment and had opportunities to antennate and exchange food with colony bees; while the control bees had normal, unrestricted access to other colony members. In all five replicates, bees reared in double-screen cages showed significantly higher levels of JH than control bees. When tested in a colony, the behavior of these bees also was consistent with their JH profile: bees reared in double-screen cages were more likely to become precocious foragers than control bees. Bees reared in single-screen cages, however, are only partially inhibited, both hormonally and behaviorally. These results suggest that direct physical contact is required for the inhibition, and that chemicals produced in the mandibular glands are passed around during either antennation or exchange of food.

VI. Recent findings and the activator-inhibitor model

Some recent findings suggest that the regulation of division of labor in honey bee colonies is a more complex process than our model has suggested. In the following I briefly discuss these findings and their implication to the model.

1. Is JH the activator?

When the model was developed, JH was the ideal candidate as the activator. Besides the correlation between high JH titers and foraging behavior, JH treatment also causes bees to forage earlier. It is therefore reasonable to assume that workers need to have certain JH levels before they can embark on the foraging career. Several lines of evidence now suggest that this is not true. First, workers are able to forage with low JH titers or rates of biosynthesis in late fall or early spring. Sometimes these foragers had JH levels as low as nurses in summer (below 2 picomole per hour per CA for rates of synthesis, or below 100 nanogram per ml for titer), despite the fact that they were sampled while returning from foraging (Huang and Robinson 1995). This is puzzling because if bees can forage with low JH levels, why do foragers in summer always have high JH levels? A second line of evidence came from experiments removing the source of JH in workers. If JH is the activator, one would assume that if the source of JH is removed, then workers perhaps would become "Peter Pan" bees and never grow up to be foragers. When allatotectomy, a surgery that removes the corpora allata, is performed on workers, they are still able to forage, but at a significantly later age (Sullivan et al. 1996). This delay can be reversed by applying a JH analog. JH therefore seems to be involved in the timing of foraging, but is not absolutely necessary for the foraging behavior.

It is possible that JH is normally highly correlated with another chemical (let us call it X). High X induces workers to become foragers, but high X bees do not always maintain high JH levels. This suggestion is consistent with the finding of JH levels in middle aged bees (Huang et al. 1994). Middle aged workers can perform several tasks: comb building, food storing, guarding and undertaking. We found that low JH levels are associated the first two tasks, and high levels with the last two tasks. There must be other factor(s), in addition to JH, to regulate division of labor in middle aged bees, because middle aged bees with high JH still have to "decide" between guarding and undertaking.

2. Is there more than one activator or inhibitor?

Our model presumed a single activator and inhibitor, but evidence is already accumulating that this might not be correct. Besides the chemical activator, be it JH or chemical X, behavioral signals might also be involved in regulating behavioral development. A recent study (Seeley *et al.* 1996) indicates that the tremble dance, a behavior performed by nectar foragers when they find it difficult to unload their harvest, can speed up behavioral development in younger bees. When the number of bees performing the tremble dances was increased artificially using an artificial feeder, the mean age for food storers changed from 24.9 days to 19.6 days in one experiment. However, it is not known if the bees that became food storers earlier also foraged earlier.

In addition to the worker-originated inhibitor, two other chemical complexes are also known to affect behavioral development in workers. Queen mandibular pheromone has been shown to inhibit JH biosynthesis in workers in cages or small colonies (Kaatz *et al.* 1992). In addition, queen pheromone was recently shown to delay age of first foraging in workers from large colonies (Pankiw *et al.* 1988). Brood pheromone also has a similar inhibitory effect on JH titer and age of first foraging in both caged bees and triple cohort colonies (Le Conte and G.E. Robinson, unpublished results).

V. Conclusions

The activator-inhibitor model is a simple model that makes testable predictions regarding behavioral development in worker honey bees. Regulation of division of labor in honey bees is undoubtedly much more complex than our model would suggest. However, good models always try to simplify because a model including everything would be too complicated to analyze (Maynard Smith 1972). I believe that the activator-inhibitor model provides a framework for further experiments and helps us understand the intricate mechanisms of social integration in honey bee colonies (e.g. Schulz *et al.* 1998, Giray *et al.* in preparation). As our knowledge about honey bee division of labor advances, the model undoubtedly will be modified and improved, but it remains a useful guide for understanding the process of division of labor in honey bee colonies.

Appendix: methods for measuring JH levels

1. Measurement of rates of JH biosynthesis

This method measures how much JH is produced by isolated corpora allata (CA). It is accepted that in most insects the isolated CA incubated under proper conditions (in vitro) would produce amount of JH similar to when they are inside the body (in vivo). The method involves 4 steps: dissection, incubation, extraction and quantification. CA are dissected from cold anesthetized workers. Care is taken not to damage the tissue because that would reduce JH production. The dissected CA are then incubated in a medium that contains amino acids, salts and sugars, with a similar composition to that of the bee blood. After all glands are dissected (usually from 20 or 40 bees), a radioactive precursor to JH (an amino acid called methionine) is added to start the incubation. Each molecule of JH made by the CA contains one radioactive element from methionine. After 3 hours of incubation, the glands are removed from the medium and JH produced (and released into the medium) can be extracted. The extraction is based on the fact that JH is fat soluble while methionine is water soluble. To do this, an organic phase (either hexane or isooctane) is added to the tube. After vortexing and centrifugation, most radioactive JH (about 95-98%) ends up in the upper organic phase, while virtually all of the radioactive methionine remains in the aqueous phase. Because the specific radioactivity of JH produced is the same as that of the methionine, the amount of JH can be calculated from the radioactivity contained in the organic phase. Radioactivity is usually measured by a Liquid Scintillation Counter, which measures how many photons are emitted by the sample (a scintillant added to the sample absorbs the radioactive decay energy and converts that to photons). For a more technical account of this method, refer to Huang *et al.* (1991).

2. Measurement of JH titer

This method measures the number of JH molecules in a blood sample. Because one knows exactly how much blood is taken, a concentration can be calculated. It involves 5 major steps: bleeding, extraction, incubation, separation and quantification. Blood from workers are obtained by puncturing a hole in the abdomen and inserting a capillary tube. The blood is immediately mixed with an organic solvent (acetonitrile) to denature any enzymes that might degrade JH in the blood. The blood-acetonitrile mixture can be store in the freezer until extraction. JH is extracted from the blood-acetonitrile mixture by adding hexane, vortexing and centrifugation. The hexane containing the JH is dried down and the JH dissolved by adding small amounts of methanol. This is then transferred another tube, which contains a mixture of radioactive JH and an antibody to JH (obtained from rabbits which are injected with JH). The antibody is highly specific for JH so it binds very few other molecules other than JH. The number of molecules that can bind JH is limited in the mixture (of JH antibody and radioactive JH), so the JH in samples and the radioactive JH would compete for the same binding sites. If there are a lot of JH in the sample, more radioactive JH would be displaced from the binding sites on the antibody, and vice versa. After the reaction reaches an equilibrium (usually two hours of incubation), the JH that is bound to the antibody and those that are free are separated. To do this, a charcoal solution is added and the tubes are centrifuged. The charcoal absorbs all the free JH, but leaves JH that are bound to antibodies in the solution, because the antibodies are too big for charcoal particles to absorb. The charcoal settles to the bottom of the test tubes after centrifugation, and the supernatant now contains only JH that are bound to the JH antibody. The amount of the bound radioactive JH can be quantified using a Liquid Scintillation Counter. To convert the radioactivity to number of JH molecules, a standard curve with known amounts of JH is run each day the samples are run. A detailed technical summary is presented by Huang *et al.* (1994).

Acknowledgment

I thank G.E. Robinson from University of Illinois for our fruitful collaborations over the last several years, B. Stay from University of Iowa for teaching me the radiochemical assay, D.W. Borst from Illinois State University for teaching me the radioimmunoassay and providing the antiserum for JH, Jack Kuehn for providing assistance in bee colony maintenance. G. Bloch, E.A. Capaldi, D.J. Shulz, D.P. Toma and M.J. Vermiglio provided comments that greatly improved the manuscript. Research reported here was supported by grants from the National Institute of Health, the National Science Foundation and the United States Department of Agriculture to G.E. Robinson. Support was provided to me by USDA (#97-35302-4784) while writing this review.

References

Breed, M.D., G.E. Robinson and R.E. Page. 1990. Division of labor during honey bee colony defense. *Behav. Ecol. Sociobiol.* 27: 395-401.

Fahrbach, S.E. and G.E. Robinson. 1996. Juvenile hormone, behavioral maturation, and brain structure in the honey bee. *Dev. Neurosci.* 18:102-114.

Fluri, P., M. Lüscher, H. Wille and L. Gerig. 1982. Changes in weight of the pharyngeal gland and haemolymph titres of juvenile hormone, protein and vitellogenin in worker honey bees. *J. Insect Physiol.* 28: 61-68.

Franks, N.R. and C. Tofts. 1994. Foraging for work: how tasks allocate workers. *Anim. Behav.* 48: 470-472.

Fukuda, H. 1960. Some observations on the pollen foraging activities of the honey bee, *Apis mellifera* L. *J. Fac. Sci.* Hokkaido Univ. Ser 6. 14: 381-386.

Giray, T. and G.E. Robinson 1994. Effects of intracolony variability in behavioral development on plasticity of division of labor in honey bee colonies. *Behav. Ecol. Sociobiol.* 35: 13-20.

Giray, T., Z.-Y. Huang, E. Guzmán-Novoa and G.E. Robinson. Physiological bases of genetic variation for rate of behavioral development in the honey bee, *Apis mellifera.* in preparation.

Hölldobler, B. and E.O. Wilson. 1990. The Ants. Belknap/Harvard University Press, Cambridge, MA.

Huang, Z.-Y. and Otis G.W. 1991. Inspection and feeding of larvae by worker honey bees (Hymenoptera: Apidae): Effect of starvation and food quantity. *J. of Insect Behav.* 4: 305-317.

Huang, Z.-Y. and G.E. Robinson. 1992. Honeybee colony integration: worker-worker interactions mediate hormonally regulated plasticity in division of labor. *Proc. Natl. Acad.Sci.* USA 89: 11726-11729.

Huang, Z.-Y. and G.E. Robinson. 1995. Seasonal changes in juvenile hormone titers and rates of biosynthesis in honey bees. *J. Comp. Physiol.* B 165: 18-28.

Huang, Z.-Y. and G.E. Robinson. 1996. Regulation of honey bee division of labor by colony age demography. *Behav. Eco. Sociobiol.* 39: 147-158.

Huang, Z.-Y., E. Plettner and G.E. Robinson. 1998. Effects of social environment and worker mandibular glands on endocrine-mediated behavioral development in honey bees. *J. Comp. Physiol.* A. 183: 143-152.

Huang, Z.-Y., G.E. Robinson and D.W. Borst. 1994. Physiological correlates of division of labor among similarly aged honey bees. *J. Comp. Physiol.* A 174: 731-739.

Huang, Z.-Y., G.E. Robinson, S.S. Tobe, K.J. Yagi, C. Strambi, A. Strambi and B. Stay. 1991. Hormonal regulation of behavioural development in the honey bee is based on changes in the rate of juvenile hormone biosynthesis. *J. Insect Physiol.*37: 733-741.

Jaycox, E.R. 1976. Behavioral changes in worker honey bees (*Apis mellifera* L.) after injection with synthetic juvenile hormone (Hymenoptera: Apidae). *J. Kans. Entomol. Soc.* 49: 165-170.

Jaycox, E.R., W. Skowronek, G. Gwynn. 1974. Behavioral changes in worker honey bees (*Apis mellifera*) induced by injections of a juvenile hormone mimic. *Ann. Entomol. Soc. Am.* 67: 529-34.

Jung-Hoffman, I. 1966. Die Determination von Königin und Arbeiterin der Honigbiene (*Apis mellifera* L.). *Bienenforsch* 8: 296-322.

Kaatz, H.-H., H. Hildebrandt and W. Engels. 1992. Primer effect of queen pheromone on juvenile hormone biosynthesis in adult worker honey bees. *J. Comp. Physiol.* 162: 588-592.

Kolmes, S.A., M.L. Winston and L.A. Fergusson. 1989. The division of labor among worker honey bees (Hymenoptera: Apidae): The effects of multiple patrilines. *J. Kansas Ent.Soc.* 62: 80-95.

Maynard Smith, J. 1972. On Evolution. University Press, Edinburgh.

Michner, C.D. 1974. The Social Behavior of the Bees: a Comparative Study. Belknap/Harvard University Press, Camrbidge, MA.

Milojévic, B.D. 1940. A new interpretation of the social life of the honeybee. *Bee World* 21: 39-41.

Nelson, F.C. 1927. Adaptability of young bees under adverse conditions. *Am. Bee J.* 67: 242-243.

Naumann, K. and M.L. Winston. 1990. Effects of package production on temporal caste polyethism in the honey bee (Hymenoptera: Apidae). Ann.Entomol. Soc. Am. 83: 264-270.

Oster, G.F. and E.O. Wilson. 1978. Caste and Ecology in the Social Insects. Princeton University Press, Princeton, NJ.

Page, R.E., Jr, G.E. Robinson, D.S. Britton and M.K. Fondrk. 1992. Genotypic variability for rates of behavioral development in worker honeybees (*Apis mellifera* L.). *Behav. Ecol.* 3: 173-180.

Pankiw, T., Z.-Y. Huang, M.L. Winston and G.E. Robinson. 1998. Queen mandibular gland pheromone influences worker honey bee (*Apis mellifera* L.) foraging ontogeny and juvenile hormone titers. *J. Insect Physiol.* 44: 685-692.

Ribbands, C.R. 1952. Division of labour in the honeybee community. Proc Roy Soc London B 14: 32-42.

Robinson, G.E.. 1985. Effects of a juvenile hormone analogue on honey bee foraging behaviour and alarm pheromone production. *J. Insect Physiol.* 31: 277-282.

Robinson, G.E.. 1987. Regulation of honey bee age polyethism by juvenile hormone. Behav Ecol Sociobiol 20: 329-338.

Robinson, G.E.. 1992. Regulation of division of labor in insect societies. *Annu. Rev. Entomol.* 37: 637-665.

Robinson, G.E. and F.L.W. Ratnieks. 1987. Induction of premature honey bee (Hymenoptera: Apidae) flight by juvenile hormone analogs administered orally or topically. *J. Econ.Entomol.* 80: 784-787.

Robinson RE, and E.I. Vargo. 1997. Juvenile hormone in adult eusocial Hymenoptera: gonadotropin and behavioral pacemaker. *Arch. Insect Biochem. Physiol.* 35: 559-583.

Robinson, G.E., A. Strambi, C. Strambi, Z.L. Paulino-Simões, S.O. Tozeto, J.M.N. Barbosa. 1987. Juvenile hormone titers in Africanized and European honey bees in Brazil. *Gen. Comp. Endocrinol.* 66: 457-459.

Robinson, G.E., R.E. Page, C. Strambi and A. Strambi. 1989. Hormonal and genetic control of behavioral integration in honey bee colonies. *Science* 246: 109-112.

Robinson, G.E., R.E. Page, C. Strambi and A. Strambi. 1992. Colony integration in honey bees: mechanisms of behavioural reversion. *Ethology* 90: 336-350.

Robinson, G.E., R.E. Page and Z.-Y. Huang. 1994. Temporal polyethism in social insects is a developmental process. *Anim. Behav.* 48: 467-469.

Rutz, W., L. Gerig, H. Wille and M. Lüscher. 1976. The function of juvenile hormone in adult worker honeybees, *Apis mellifera. J. Insect Physiol.* 22: 1485-1490.

Sasagawa, H., M. Sasaki and I. Okada. 1989. Hormonal control of the division of labor in adult honeybees (Apis mellifera L.) I. Effect of methoprene on corpora allata and hypopharyngeal gland, and its a-glucosidase activity. *Appl. Ent. Zool.* 24: 66-77.

Schulz, D.J., Z.-Y. Huang and G.E. Robinson. 1998. Effect of colony food shortage on the behavioral development of the honey bee, *Apis mellifera. Behav. Ecol. Sociobiol.*; 42: 295-303.

Seeley, T.D., S. Kühnholz and A. Wiedenmuller. 1996. The honey bee's tremble dance stimulates additional bees to function as nectar receivers. *Behav. Ecol. Sociobiol.* 39: 419-427.

Simpson, J., I.B.M. Riedel and N. Wilding. 1968. Invertase in the hypopharyngeal glands of the honey bee. *J Apic. Res* .7: 29-36.

Strambi C., A. Strambi , M.L., de Reggi and M.A. Delaage. 1984. Radioimmunoassays of juvenile hormones: State of the methods and recent data on validation. In: Biosynthesis, Metabolism and Mode of Action of Invertebrate Hormones (Edited by Hoffmann J. and Porchet M.). pp. 356-332. Springer-Verlag, Berlin.

Sullivan, J.P., O. Jassim, G.E. Robinson and S.E. Fahrbach. 1996. Foraging behavior and mushroom bodies in allatectomized honey bees. *Soc. Neuro. Abst.* 22: 1144.

Tofts, C., N.R. Franks. 1992. Doing the right thing: ants, honeybees and naked mole-rats. *Trends Ecol. Evol.* 7: 346-349.

Wilson, E.O. 1985. The sociogenesis of insect colonies. *Science* 228: 1489-1495.

Winston, M.L. 1987. The Biology of the Honey Bee. Harvard Univ. Press, Cambridge, MA

Winston, M.L. and L.A. Fergusson. 1985. The effect of worker loss on temporal caste structure in colonies of the honeybee (*Apis mellifera* L.). *Can. J. Zool.* 63: 777-780.

Winston, M.L. and L.A. Fergusson. 1986. The influence of the amount of eggs and larvae in honey bee (*Apis mellifera* L.) colonies on temporal division of labour. *J. Apic. Res.* 25: 238-241.

Winston, M.L. and E.N. Punnett. 1982. Factors determining temporal division of labour in honeybee. *Can. J. Zool.* 60: 2947-2952.

Keywords: Activator-inhibitor model, age polyethism, *Apis mellifera*, division of labor, juvenile hormone, JH biosynthesis, JH titer, queen mandibular pheromone, behavioral development, social inhibition, social interaction.

Displacement of European honey bee subspecies by an invading African subspecies in the Americas

ORLEY R. "CHIP" TAYLOR
DIVISION OF BIOLOGICAL SCIENCES,
DEPARTMENT OF ENTOMOLOGY,
UNIVERSITY OF KANSAS
LAWRENCE, KS 66045

Phone: 913-864-4051
Fax: 913-864-4441
Email: monarch@ukans.edu

Abstract

The introduction of a subspecies of the common honey bee, *Apis mellifera*, from South Africa into Brazil in 1956 resulted in a spectacular biological invasion which has had profound effects on agriculture, human and animal health and the science of honey bee genetics. The introduced subspecies, *Apis mellifera scutellata*, escaped or was intentionally released from a study site in southern Brazil in early 1957. A large population of feral bee colonies soon became established and began to spread at rates of 100-300 miles per year. The arrival of the bees in each new area was followed by numerous stinging incidents and the deaths of humans and farm animals. Thus the bees became known as "killer bees" in the press and given names such as Brazilian bees, Africanized bees or African bees by various experts. Establishment of the African bees in each location led to hybridization with the bees of European subspecies maintained by beekeepers in apiaries for honey production and pollination. The apiary bees soon became extremely defensive and difficult to manage, and honey production declined as beekeepers abandoned beekeeping.

The invasion by African bees raised many biological questions. Why were they able to invade new habitats so readily? How far would they spread in the Americas? How were they spreading so rapidly? Was this a human assisted invasion or were the bees accomplishing this colonization of new areas by natural means? Why were the European subspecies disappearing after the arrival of African bees and why, if hybridization was common, did African bees appear to be unchanged by the hybridization—in other words, why weren't the African bees becoming Europeanized? Our research over the last 22 years has focused on answering these and many other related questions. Over this period, our studies have ranged from examinations of adaptations such as reproduction and survival of feral African bee colonies, mating behavior, population genetics and the genetics of hybridization.

African bees reached the United States in 1990 and have spread slowly in Texas, New Mexico, Arizona and California. In many areas they appear to have reached their climatic limits. Genetically, the African bees in the US are little changed from those in South Africa, but virtually all genetic traces of European bees have disappeared in areas occupied by African bees for 10 years or more. There appears to have been complete genetic displacement of one biotype by another. Why and how? The African and European bees are supposed to be the same species and should therefore be completely genetically compatible. Our studies show that there are a number of barriers to gene flow from European bees into the African bee population but that the converse is not true, European bees are vulnerable to hybridization with African bees. This asymmetry in hybridization combined with low survival of the hybrid bees evidently leads to a rapid elimination of European bees and their genetic traits. The next question is what will be the nature and dynamics of the zone of contact or hybrid zone formed by these populations in the United States?

Introduction

Honey bees (*Apis mellifera*) are native to the Old World and all honey bees present in the Americas are descendants of bees introduced from Europe, Africa and the Middle East (Pellet 1938, Watkins 1968, Oertel 1976, Sheppard 1989). Managed honey bee colonies in the Americas are derived from at least five introduced European subspecies belonging to two major lineages of *A. mellifera* (west European bees, *A. m. mellifera* and *iberica*; eastern European bees, *A. m. ligustica, carnica* and *caucasica*; Smith 1991a,c). These temperate subspecies are notably successful in other temperate regions (*e.g.,* Australia) and have been used with some success in tropical areas. However, they are not well-adapted to tropical conditions and have not established large self-sustaining feral populations in these regions as they have in subtropical and temperate regions. In many tropical areas European bees would not persist without human assistance. *A. m. scutellata*, a subspecies belonging to the African lineage of *A. mellifera*, was intentionally introduced to Brazil in 1956 (Kerr 1967) in order to create a domesticated European/African hybrid that was better adapted to tropical conditions. Within a short time a feral, African-derived population that was extremely well-adapted to tropical conditions became established in Brazil (Taylor 1985) and subsequently spread over most of tropical America (Figure 1). Key differences in the biology of European- and African-derived bees that influence the outcome of the interactions between these populations are summarized in Table 1.

Table 1. Differences in key traits affecting survival of African and European bees throughout the contact zone.

Traits	African	European	Net Advantage
Overwintering	Poor	Good	European advantage which increases with duration of and severity of winter conditions (Kerr et al. 1982).
A. Honey Storage*	Low	High	European advantage (Rinderer et al. 1984; Taylor and Spivak 1984)
B. Nest size	Small-Medium	Medium-Large	European advantage (Seeley and Morse 1976; Winston and Taylor 1980; Taylor and Spivak 1984)
C. Longevity of bees	Short	Long	European advantage (Woyke 1973)
Resistance to Varroa	Good	Poor or None	Strong Advantage to Africans (Moretto et al. 1991)
Reproductive rate	High	Low	Strong advantage to Africans (Winston 1979; Otis 1980, 1982)
Assortative mating	Higher	Lower	African like-to-like matings help maintain genetic integrity (Kerry and Bueno 1970)
Queen and worker development time	Faster	Slower	In mixed matings, African paternal lines contribute to faster development of queens (Figure 4)
Absconding	high	Low	African advantage during warm months but a disadvantageeous trait in late fall or early spring (Winston, Otis and Taylor 1979)

*Significant components of overwintering success.

When two formerly isolated species or populations—such as European and African honey bees—come into secondary contact there are four possible outcomes: 1) coexistence with complete reproductive isolation, 2) replacement of one population by the other, 3) fusion of the two populations and complete mixing of the two gene pools (some times referred to as "dilution"); and 4) establishment of a more or less permanent hybrid zone.

The evidence of the past forty years shows that the first scenario has not occurred. In the neotropics the second scenario seems to be the rule in non-managed populations: African-derived bees establish large feral populations, and replace any resident European feral honey bees (Michener 1975; Spivak 1991, Taylor 1985, 1988; Taylor *et al.* 1991). The pattern of gene flow between managed European populations and feral African-derived populations in the neotropics has been the subject of a great deal of research.

Gene flow

Africanization of European-derived bees in managed apiaries occurs mainly through mating of European queens with African-derived drones from the feral population (Taylor 1985). Thus, gene flow from the feral African-derived population to European apiary populations is primarily via males, through take-over of apiary colonies by African swarms or queens has been documented (Michener 1975, Rinderer *et al.* 1991, Vergara *et al.* 1989).

The feral African-derived population, on the other hand, could experience gene flow from the European apiary population 1) via males, if African queens mate with European males; or 2) via females, if pure or hybrid (European mother/African father) queens disperse from apiaries in swarms and join the feral population.

Studies by Smith *et al.* (1989) and Hall and Muralidharan (1989) showed that European mitochondrial DNA (mtDNA) is much rarer than African mtDNA in the feral African-derived population. Of 55 feral swarms collected in Tapachula, Mexico after the arrival of African-derived bees in the area, 54 had African mtDNA. Samples from 34 of 35 colonies from Brazil (10), Venezuela (22) and Costa Rica (3) also had African mtDNA. Results from our recent study in northern Mexico show a rapid decline in the frequency of swarms with European mtDNA. The study was initiated 3 months after African bees were first found in the area. Four of the

Figure 1. Spread of introduced African honey bees in the Americas from 1957 to 1990. The 10°C line in Argentina represents the mean temperature for the coldest month. Modified from Thomas (1991).

first five swarms collected had European mtDNA but the proportion of European mtDNA for the entire year was only .26 (n=216). In the third year, the proportion had declined to .06 (n=90) (Smith and Taylor, unpublished). The rate of decline in European mtDNA is evidently slower in areas, such as the Yucatan Peninsula, which have high densities of managed European bees. Recently, Quezada-Euan and Hinsull 1995 found .20 (n=65) European mtDNA. Overall, this pattern indicates that the feral African-derived population is NOT experiencing extensive gene flow from the European apiary population via females. If pure or hybrid (European mother/African father) queens dispersed from apiaries and persisted in the feral population, we would expect to find European mtDNA in high frequency in the feral population.

The other avenue for gene flow from the European apiaries to the feral African-derived population is via males. Studies of mtDNA provide no information on this point. However, studies of allozymes (Smith *et al.* 1989) and nuclear restriction fragment length polymorphisms (Hall and Muralidharan 1989) indicate that the feral African population maintains African-specific markers in high frequency despite initial hybridization with Eu-

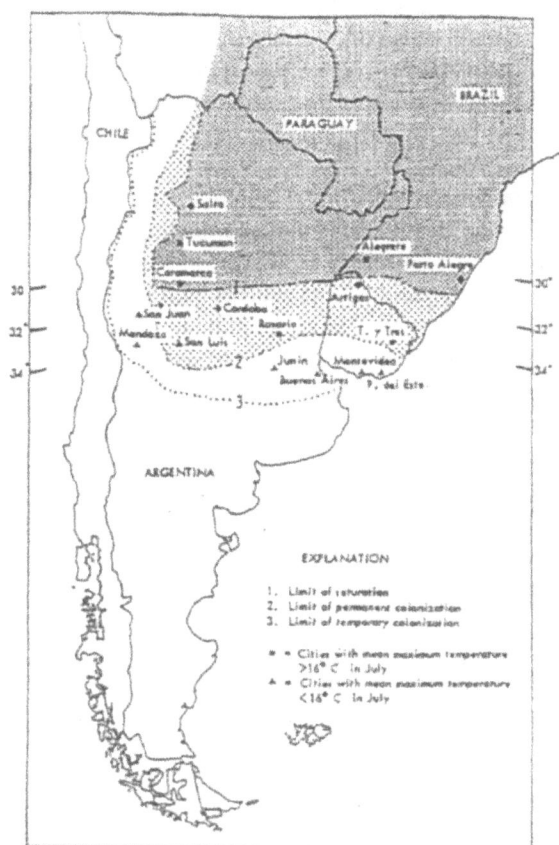

Figure 2. Distribution of neo-tropical honey bees in South America. The areas between lines 1 and 2 define a permanent hybrid zone in which both parental types and numerous hybrid combinations have been found (Sheppard et al. 1991). The areas between 2 and 3 define regions which African bees migrate into during warmer months. The data supporting these zones is sketchy and the population dynamics and patchiness within the zones has not been studied. The data by Sheppard et al. (1990) can be interpreted to indicate that either African mtDNA is introgressing into European bee populations to the south or the zone of contact is broader than indicated (from Kerr et al. 1982).

ropean bees. Boreham and Roubik (1987) collected samples of feral honey bees in Panama as African bees were invading the region. They found that the earliest samples of bees were larger than typical South African *A. m. scutellata*, and larger than feral African-derived bees from populations further south. This implies hybridization with the larger European honey bees. However within 3 years the size of feral honey bees in Panama decreased to a size similar to or even smaller than *A. m. scutellata*.

This implied that either the smallest bees were out-reproducing larger bees, or that some form of selection removed larger bees from the population. Similarly Hall (1990, 1992ab), using European- and African-specific nuclear RFLPs finds less evidence of Europeanization of African bee samples recently obtained from a location in Venezuela, which the feral African population first reached in 1978, than from samples obtained in Mexico in the early stages of invasion by African bees. Data from our studies in Mexico on two allozyme markers, hexokinase (HK) and malae dehydrogenase (MDH), suggest that European traits acquired in the early phase of hybridization are rapidly lost from the feral population. Remarkably, feral populations of African-derived bees are converging on similar MDH and HK allele frequencies from southern Brazil (Del Lama *et al.* 1988, Lobo *et al.* 1989; Sheppard *et al.* 1991) to northern Mexico.

The unanswered question is: Why do we observe these patterns? Selection may favor African-derived over pure European bees in the neotropics, but this cannot be the whole answer. In highly Africanized apiaries, where queens have mated with African drones for many generations, queens will still have European mtDNA, but a largely African nuclear genome. If the sole factors involved in the success of the feral African population involved traits coded in the nuclear genome, the numerous swarms produced by these queens should be successful in feral tropical populations and should introduce European mtDNA into the African-derived feral population. Yet we don't find European mtDNA in colonies with African allozyme patterns or morphometric characteristics. This suggests that there may be barriers to introgression and selection against hybrids, perhaps via interactions between nuclear and cytoplasmic factors (Harrison and Hall 1993).

Barriers to gene flow
Assortative Mating
Assortative mating and queen developmental rates may also play a role in maintaining a high frequency of African genetic markers in the feral population. Paternity analysis of the progeny from African-derived ($n=34$) and European-derived queens ($n=49$) raised in, and mating from, the same apiaries show that African-derived queens mated with few (0-6%) European drones in areas where European drones were abundant. In contrast, the European queens mated with 20-35% European

drones and 65-80% African drones. In the absence of European drones, both types of queens mated with the expected genotypic frequencies for feral (African) drones. The result obtained for the African queens show that even in the presence of large numbers of European drones, matings with these drones were few, suggestive of some form of pre reproductive isolation.

Queen Development

Data on developmental times of European- and African-derived queens and F_1 hybrid queens show that queens fathered by African drones develop faster than those fathered by European drones (Figure 4). The first queen to emerge usually becomes the next queen to head the colony. In cases where European or African queens mate with both types of drones the next queen is likely to be hybrid rather than pure European in a European colony, and pure African rather than hybrid in an African colony. The effect of these differences in development could be to increase introgression of African traits into the European populations, and decrease, European traits in the African population.

Hybrid Breakdown

Evidence from studies of respiration and honey production suggest hybrid dysfunction or breakdown is important in this system. Harrison and Hall (1993) have measured mass-specific metabolic capacities of worker honey bees from "African", "European" and hybrid colonies. Feral African-derived colonies from Honduras were used for the African population, Hawaiian Italian-derived colonies were used for the European population, and crosses were made by instrumental insemination. They found that the offspring of African mothers and African fathers had the highest mass-specific metabolic

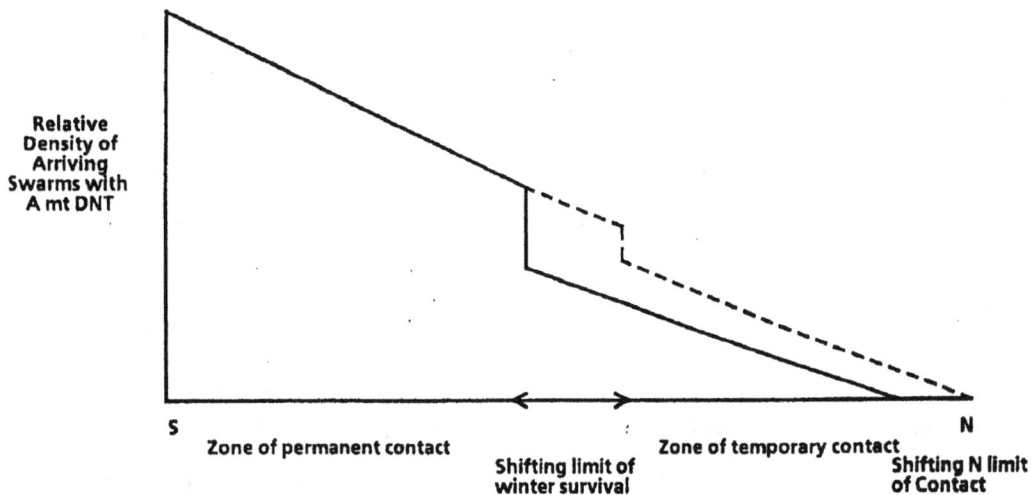

Figure 3. Relative density of arriving swarms with African mt DNA throughout the zone of overlap. Two subzones will be formed: a zone of permanent contact in which both parental types and numerous hybrid combinations well persist due to yearly production of reproductive swarms from apiaries with European mt DNA and continuous (warm season) dispersal of feral swarms (reproductive and absconding) with African mt DNA into the zones and a subzone of temporary contact, situated north of the overwintering limit, into which swarms with African mt DNA will disperse during the warmer months. The limits of overwintering will shift South or North depending on the severity of the winter. The extent of dispersal beyond the overwintering limit will be a function of the size of the feral population surviving the winter south of the wintering limit and the degree to which spring conditions favor swarming and dispersal. In this study we will define the zone of permanent contact in central Texas and within this zone we will establish patterns of introgression of several selectively advantageous and neutral traits associated with African or European bees into the opposite populations. This introgression will occur against an extremely heterogenous biotic background with extreme spatial and temporal populations. This introgression will occur against an extremely heterogenous biotic background with extreme spatial and temporal variability in the density of European bee populations, floral resources and nesting sites for feral colonies, and parasitic mites.

Taylor: Displacement of European bees by African subspecies

Asymmetries in queen development

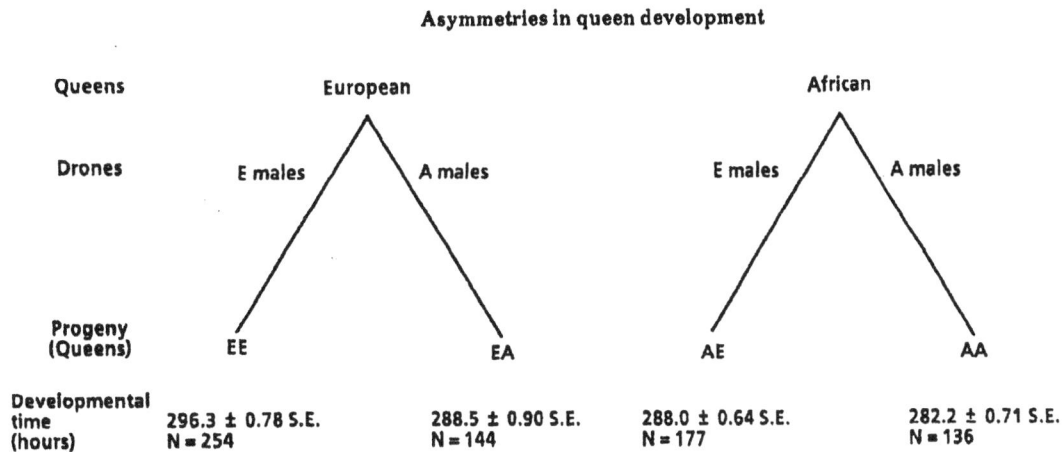

Figure 4. Asymmetries in queen development may drive "Africanization" of European bees and lead to the loss of European traits in African populations. In honey bees, during queen replacement (swarming and supersedure), the first queen to emerge usually becomes the next queen to head the colony. When European queens mate with both European and African drones, the hybrid (EA) progeny (queens) have an 8 hour developmental advantage. Conversely, when African virgin queens mate with both kinds of drones, the African queen progeny have a developmental advantage of almost 6 hours. Daughter queens reared from a European queen inseminated with semen from 4 European and 1 African drone conform to these expectations. The mean developmental times of EA progeny (N=17) were 6 hours shorter than those of EE (N=77) progeny, but more importantly, in each replicate (N=3) an EA queen emerged first. Overall, 8 of the first queens to emerge were EA.

capacities (measured as watts per kg thorax tissue), and the offspring of European mothers and European fathers had substantially lower metabolic capacities. F_1 crosses and backcross hybrids had metabolic capacities as low or lower than those of the "pure" Europeans. In addition, metabolic rates of hybrid progeny of European queens x African drones were significantly lower than those of progeny of African queens x European drones. These results suggest that hybrids may be less fit than parentals, perhaps because of disruption of co-adapted enzyme complexes, and that nuclear/cytoplasmic incompatibility may put the progeny of European queens x African drones at a particular disadvantage. In subsequent tests, Hall and his students found lower brood production rates (Hall, pers. com.) and lower honey production in backcross hybrids (Villavicencio *et al.* 1993). Since brood and honey production are attributes which are important for survival, the implication is that the fitness of backcrosses is lower than that of parentals and F1s.

Hybrid zones

The interactions between the European and African populations change as African bees advance into subtropical and temperate regions, where Eu-

ropean bees are expected to be more fit as a feral population (Taylor 1985). Under these conditions, the outcome of contact between African- and European-derived populations could be fusion of the two populations and complete mixing of the two gene pools (the third of the scenarios listed above) or establishment of a more or less permanent hybrid zone (fourth scenario) (Figure 3).

The latter has occurred in temperate Argentina. African-derived bees in Argentina have extended their range south to approximately the latitude of Buenos Aires (Kerr *et al.* 1982). A hybrid zone exists in this region (Figure 2), with predominantly African feral populations to the north, predominantly European (western European) populations to the south, and a 'hybrid swarm' between them. This hybrid population contains bees with many combinations of European and African morphometric characteristics and mitochondrial genomes (Sheppard *et al.* 1991).

Hybrid zones are a common feature of range contacts between closely related allopatric taxa (*e.g.*, Endler 1977). Within a hybrid zone individuals with hybrid ancestry are relatively abundant; outside the zone such individuals are rare in both parental populations. Hybrid zones are of great interest both for what they can reveal about causes of a particular taxon's range limits, and as "laboratories" where natural genetic experiments occur as elements of the two parental genomes are recombined and tested by selection (Hewitt 1988). In general the origin of existing hybrid zones cannot be determined from the characteristics of the zone itself (Barton and Hewitt 1995, Endler 1977). Geological, biogeographical, and historical evidence (if available) must be used to infer the events that led to establishment of the zone. In the case of a hybrid zone between European-derived and African-derived honey bees in the Americas, we have an unambiguous case of secondary contact after differentiation in geographic isolation.

Theoretical and empirical studies of other hybrid zones (Barton 1983, Barton and Hewitt 1985) have suggested several processes potentially responsible for maintenance of stable zones including: (1) clinal selection against original (parental) alleles (Endler 1977) and (2) epistatic interactions among loci (Arnold *et al.* 1988, Asmussen *et al.* 1987, 1989, Barton 1979, 1983, Szymura and Barton 1986). Epistatic interactions among loci can be either deleterious, making the hybrid less fit relative to parental types (*e.g.*, Szymura and Barton 1986, Hagen 1990), or advantageous, leading to hybrids with high relative fitness at the hybrid zone (Moore 1977).

Evidence on the forces responsible for maintaining a hybrid zone can be obtained through estimates of cline positions and widths for independent loci, and the magnitude of linkage disequilibria among loci (Barton 1979, 1983, Barton and Hewitt 1985, Mallet *et al.* 1990). Hybrid zones maintained by purely epistatic interactions ("tension zones," Barton and Hewitt 1985) are free to move and may become hyped by local barriers or regions of low population density. Spread of neutral alleles across such zones may be impeded (Barton 1979, 1983). In contrast, hybrid zones maintained by selection gradients may be more permeable to neutral alleles. In these hybrid zones the positions of clines for different loci may not coincide.

We expect a hybrid zone to be established in the temperate United States as has occurred in Argentina. Projected overwintering limits of feral African-derived honey bees in Texas (Taylor and Spivak 1984, Taylor 1985) are based on temperature isoclines during the coldest month at the distributional limits of African-derived bees in Argentina (Kerr *et al.* 1982) and of *A. m. scutellata* in South Africa (Taylor 1977). We predict that northward movement of the African-derived population will stop in central Texas, and that a zone of hybridization will be established, with a predominantly European honey bee population to the north and a predominantly neotropical African population to the south.

Projections by Southwick *et al.* (1990) and by Deitz *et al.* (1986) place the overwintering limits much farther north but both fail to consider that African-derived bees (1) are short lived as overwintering adults (Woyke 1973), and thus may not live long enough to survive a long period during which no new brood can be reared, and (2) do not survive in either South Africa or Argentina where "windows" (as defined by first and last frost, or by duration of near zero availability of nectar and pollen) are more than 3.5 months in duration (Taylor and Spivak 1984, Woyke 1973).

Experimental studies can be used to test explicit hypotheses about selective factors. In the present case, there is evidence for both clinal selection favoring parental types (*e.g.*, selection on overwintering ability, *Varroa* resistance) and evidence for negative epistatic interactions in backcross hybrids (*e.g.*, lower mass-specific metabolic rates, lower honey production). Presently, we are studying the spatial and temporal dynamics and population genetics of the hybrid zone which is forming between African and European bees in Texas.

Conclusions

Gene flow between neotropical African and European bees appears to be strongly asymmetrical. African bees have maintained their genetic integrity, in spite of hybridizing with European bees, as they have expanded their distribution over the last 40 years. Low acquisition of European traits into the African population can be attributed to pre and post zygotic isolating mechanisms, *i.e.*, mate selection, queen developmental time and hybrid dysfunction. European bees become rapidly Africanized and nearly all traces of the nuclear and mitochondrial genome disappear from the feral bee populations following the arrival of African bees. The disappearance of the European traits seems to be due to a lack of pre reproductive isolation which results in extensive mating by European queens with African drones. This is followed by a pattern of queen development which favors hybrid rather than European queens. Matings by these F_1 queens to African drones results in colonies with low fitness and the eventual loss of European mtDNA from the population. Displacement of European bees therefore seems to be due, in part, to a type of "genetic capture" in which one form, *A. m. scutellata*, eliminates the others by hybridizing with their females. This type of displacement may not be uncommon, a similar case is known for sulfur butterflies (*Colias*) (Grula and Taylor 1980). The genetic and population consequences of the interactions between *A. m. scutellata* and *A. mellifera* subspecies from Europe suggest that *A. m. scutellata* deserves the status of a semi-species.

Acknowledgments

I wish to thank Felipe Brizuela, J. Christopher Brown, Anthony Delgado, Michael Engle, Gordon Johnston, Andrew Jong, David McMahon, Nancy Oderkirk, Elizabeth Smith for their assistance in the field as well as the laboratory and H. Glenn Hall, Gard Otis, Deborah Smith, Marla Spivak, Jose Villa and Mark Winston for formative discussions on the adaptations of African bees.

Literature cited

Arnold, J., Asmussen, M. A., and Avise, J. C. 1988. An epistatic mating system model can produce permanent cytonuclear disequilibria in a hybrid zone. *Proc. Natl. Acad. Sci. USA* 85:1893-1896.

Asmussen, M. A., Arnold, J. and Avise, J. C. 1989. The effects of assortative mating and migration on cytonuclear associations in hybrid zones. *Genetics* 122:923-934.

Asmussen, M. A., Arnold, J., and Avise, J. C. 1987. Definition and properties of disequilibrium statistics for associations between nuclear and cytoplasmic genotypes. *Genetics* 155:755-768.

Barton, N. H. 1979. Gene flow past a cline. *Heredity* 43:333-339.

Barton, N. 1983. Multilocus clines. *Evolution* 37:454-471.

Barton, N. H. and Hewitt, G. M. 1985. Analysis of hybrid zones. *Annu. Rev. Ecol. Syst.* 16:113-148.

Boreham, M. M. and D. W. Roubik. 1987. Population change and control of African honey bees in the Panama Canal area. *Bull. Ent. Soc. Amer.* 23:34-39.

Del Lama. M. A., R. A. Figueiredo, A., E. E. Soares and S. N. Del Lama. 1988. Hexokinase polymorphism in *Apis mellifera* and its use for Africanized honeybee identification. *Rev. Brazil. Genet.* 11:287-297.

Deitz, A., Krell, R. and Pettis, J. 1986. Study on winter survival of Africanized and European honey bees in San Juan, Argentina. pp. 87-91 In: Proc Africanized honey bee symposium, Atlanta. American Farm Bureau Research Foundation, Park Ridge, IL.

Endler, J. 1977. Geographic variation, speciation and clines. Princeton University Press, Princeton NJ.

Grula, J. W. and O. R. Taylor. 1980. The effect of X-chromosome inheritance on mate-selection behavior in the sulfur butterflies, *Colias eurytheme* and *C. philodice*. *Evolution* 34: 688-695.

Hagen, R. H. 1990. Population structure and host use in hybridizing subspecies of *Papilio glaucus* (Lepidoptera: Papilionidae). *Evolution* 44:1914-1930.

Hall, H. G. 1990. Parental analysis of introgressive hybridization between African and European honeybees using nuclear DNA RFLPS. *Genetics* 125:611-621.

Hall, H. G. 1992a. DNA studies reveal processes involved in the spread of New World African honey bees. *Florida Entomologist* 75:51-59.

Hall, H. G. 1992b. Further characterization of nuclear DNA RFLP markers that distinguish African and European honeybees. *Arch. Insect Biochem. Physiol.* 19:163-175.

Hall, H. G. and K. Muralidharan. 1989. African honey bees spread as continuous maternal lineages. *Nature* 339:211-213.

Harrison. J. F. and Hall, H. G. 1993. African-European honeybee hybrids have low nonintermediate metabolic capacities. *Nature* 363:258-260.

Hewitt, G. M. 1988. Hybrid zones—natural laboratories for environmental studies. *Trends in Eco. Evol.* 3: 158-167.

Kerr, W. E. 1967. The history of the introduction of African bees to Brazil. *S. Afr. Bee J.* 39:3-5.

Kerr, W. E., Del Rio, S., De Barrionuevo, M. D., 1982. The southern limits of the distribution of the Africanized bee in South America. *Am. Bee J.* 122:196-198.

Lobo, J. A., M. A. Del Lama and M. A. Mestriner. 1989. Population differentiation and racial admixture in the Africanized honeybee (*Apis mellifera* L.). *Evolution* 43:794-802.

Mallet, J., Barton, N., Lamas M. G., Santisteban C. J., Muedas M. M., and Eeley, H. 1990. Estimates of selection and gene flow from measures of cline width and linkage disequilibrium in *Heliconius* hybrid zones. *Genetics* 124:921-936.

Michener, C. D. 1975. The Brazilian bee problem. *Annu. Rev. Entomol.* 20:399-416.

Moore, W. S. 1977. An evaluation of narrow hybrid zones in vertebrates. *Quart. Rev. Biol.* 52:263-278.

Oertel, E. 1976. Early records of honey bees in the eastern United States, parts I-IV and conclusion. *Am. Bee J.* 116: 70-71, 114, 128, 156-157, 214-215, 260-261, 290.

Pellet, F. C. 1938. *History of American Bee-Keeping.* Collegiate Press, Inc,. Ames Iowa. Chapter 1 pp. 1-85.

Quezada-Euan, J. J. and S. M. Hinsull. 1995. Evidence of continued European morphometrics and mtDNA in feral colonies of honey bees (*Apis mellifera*) from the Yucatan Peninsula, Mexico. *J. Apic. Res.* 34: 161-166.

Rinderer, T. E., Stelzer, J. A. Oldroyd, B. P., Buco, S. M. and Rubink, W.L. 1991. Hybridization between European and Africanized honey bees in the neotropical Yucatan Peninsula. *Science* 253:309-311.

Sheppard, W. S. 1989. A history of the introduction of honey bee races into the United States. *Am. Bee J.* 129:617-619, 664-666.

Sheppard, W. S., Soares, A. E. E., De Jong, D., and Shimanuki, H. 1991a. Hybrid status of honey bee populations near the historic origin of Africanization in Brazil. *Apidologie* 22:643-652

Smith D. R. 1991a. African bees in the Americas: Insights from Biogeography and Genetics. *Trends in Ecol. Evol.* 6:17-21.

Smith, D. R. 1991b. Mitochondrial DNA and honey bee biogeography. pp. 131-176 *In* Smith, D. R. (ed.), Diversity in the Genus *Apis*. Westview Press, Boulder, CO.

Smith, D. R., Taylor, O. R. and Brown, W. M. 1989. Neotropical Africanized honey bees have African mitochondrial DNA. *Nature* 339:213-215.

Southwick, E. E., Roubik, D. W. and Williams, J. M. 1990. Comparative energy balance in groups of Africanized and European honey bees: ecological implications. *Comp. Biochem. Physiol.* A 97:1-7.

Spivak, M. 1991. The Africanization process in Costa Rica. pp. 137-155 *In:* M. Spivak, D. J. C. Fletcher, and M. D. Breed, eds. The "African" honey bees. Westview Press, Boulder, CO.

Szymura, J. M. and Barton, N. H. 1986. Genetic analysis of a hybrid zone between the fire-bellied toads, *Bombinia bombinia and Bombinia variegata* near Cracow in southern Poland. *Evolution* 40:1141-1159.

Taylor, O. R. 1977. The past and possible future spread of Africanized honey bees in the Americas. *Bee World* 58:19-30.

Taylor, O. R. 1985. African bees: potential impact in the United States. *Bull. Ent. Soc. Amer.* 31:14-24.

Taylor, O. R. 1988. Ecology and economic impact of African and Africanized honey bees. pp. 29-41 *In:* G. R. Needham, R. E. Page, M. Delfinado-Baker, and C. E. Bowman (eds.), *Africanized honey bees and bee mites,* Ellis Horwood Ltd., Chichester, England.

Taylor, O. R. and M. Spivak. 1984. Climatic limits of tropical African honeybees in the Americas. *Bee World* 65:38-47.

Taylor, O. R., Delgado, A. and Brizuela, F. 1991c. Rapid loss of European traits from feral neotropical African honey bee populations in Mexico. *Am. Bee J.* 131:783-784 ABSTRACT.

Thomas, J. 1991. Map in: Texas Africanized Honeybee Advisory Committee (Eds.), Texas Africanized honeybee management plan.

Vergara, C., Dietz, A. and Perez, A. 1989. Usurpation of managed honey bee colonies by migratory swarms in Tabasco, Mexico. *Am. Bee J.* 129:824-825.

Villavicencio, R. A., Suazo, A., Hall, H. G., and Sanford, M. T. 1993. Evaluating honey production in first generation and backcross hybrids of African and European honey bees in Honduras. *Am. Bee J.* 133:869.

Watkins, L. H. 1968. Some comments on Pellett's *(sic)* history. *Am. Bee J.* 108:362-363.

Woyke, J. 1973. Experience with *Apis mellifera adansonii* in Brazil and Poland. *Apiacta* 8:115-116.

Key words: African bees, hybridization, gene flow

Genetic changes of a population of feral honey bees in the Sonoran desert of southern Arizona following the arrival of *Acarapis woodi, Varroa jacobsoni* and Africanization.

G.M. LOPER[1], J. FEWELL[2], D.R. SMITH[3], W.S. SHEPPARD[4] AND N. SCHIFF[5]

[1] USDA, ARS, CARL HAYDEN BEE RESEARCH CENTER, 2000 E. ALLEN RD., TUCSON, AZ 85719
[2] ARIZONA STATE UNIV., DEPT. OF ZOOLOGY, TEMPE, AZ 85287-1501
[3] UNIV. OF KANSAS, DEPT. OF ENTOMOLOGY, LAWRENCE, KS 66045
[4] WASHINGTON STATE UNIV., DEPT. OF ENTOMOLOGY, PULLMAN, WA 99164-6382
[5] USDA FOREST SERVICE, P.O. BOX 227, STONEVILLE, MS 38776

Introduction

Colonies of feral honey bees (*Apis mellifera*) in the United States have been considered to be even more numerous than the number of colonies managed by beekeepers; in 1992, (the most recent Agricultural Statistics available, USDA, NASS, ASB. Bull. #912) managed colonies were estimated at 3.04 million, a reduction of 490,000 colonies since 1989. Since the dispersal of the tracheal mite (*Acarapis woodi*) and the varroa mite (*Varroa jacobsoni*) across the U.S., many accounts of severe losses of both domestic and feral colonies have been published (Kraus and Page 1995, Burgett 1996 and Loper 1995a, 1996a, 1997a). In addition, Africanized honey bees (AHB) were first found in southern Arizona in the spring of 1993 (Arizona Dept. of Agriculture News Release, June 18, 1993).

Since 1988, we have conducted an ecological and genetic study of a feral population of honey bees which were not directly influenced by management decisions of beekeepers. Our previously published work reported the impact of mites on the population size of this feral population of over 220 colonies nesting in rock cavities (Loper 1995a, 1996a, 1997a). In this paper, we present changes in the genetics of this population as a result of mites and the initial phases of Africanization during 1992-1996.

Materials and methods

The feral colony study area is 60 km NE of Tucson. Nests were located in rock cavities adjacent to a major drainage in this typical Sonoran desert habitat. These nest sites were in an area extending approximately 14 km alongside a set of seasonally dry stream beds. Elevation of nest sites ranged from 670 to 1,158 m. Rainfall averages 33 cm annually but in January and February, 1993 over 50 cm of rain fell, resulting in a proliferation of flowers and honey bee swarms.

Workers from all accessible colonies were collected at the colony entrances in Nov. 1991, Mar. and Oct.-Nov. 1995, Mar. and Jun. 1996 and Mar. 1997. Other representative colonies were sampled on other dates. Swarms were caught in bait hives (Schmidt *et al.* 1989) and sampled during the spring of 1993 and 1995. All samples were stored in 70% ethanol and a sub-set at -72°C. Three types of analyses were used to distinguish honey bee races in this study: morphometric (USDAID, Rinderer *et al.*

1993) mitochondrial DNA (mt DNA) and allozymes (malate dehydrogenase and hexokinase). The morphometric method involves measurements of 25 body characters including wing length and width, wing vein lengths and angles, leg measurements and wax mirror dimensions.

Mt DNA analyses were performed using nucleases (ECORI) and ^{32}P probes on extracts from honey bees, with separation on agarose gels and visualized using autoradiography. Colonies exhibiting AHB ECORI haplotypes were further analyzed using the restriction endonuclease, Hinfl, which distinguishes *A.m. lamarckii* from *A.m. scutellata* and *A.m. intermissa* (Schiff *et al.* 1994).

The allozymes Malate dehydrogenase (Mdh) and Hexokinase (Hk) exhibit polymorphisms which help characterize AHB from European (EHB) honey bees. These enzymes were extracted from individual thoraces (either 20 or 24 bees/colony) in TRIS buffer, separated using horizontal cellulose acetate electrophoresis with visualization by appropriate stains (Sheppard 1988; Del lama *et al.* 1988; Spivak *et al.* 1988). We use the designation for Mdh isozymes as fast (Mdh^{100}), medium (Mdh^{83}) and slow (Mdh^{65}), and for HK isozymes as fast (Hk^{100}) or slow (Hk^{83}). High proportions of Mdh^{100} and any level of HK^{83} are associated with Africanization (Sheppard *et al.* 1991).

The analyses for tracheal mites were performed on 30 bees/sample using the thoracic slice method of Delfinado-Baker 1984. Bees from the samples in

alcohol were decapitated, and using a razor blade, a thin slice of the thorax was removed, cleared overnight in 5% potassium hydroxide (KOH) and the tracheal tubes examined for the presence or absence of tracheal mites. The infestation level was reported as a percent of the workers with mites in each sample. Varroa mite infestation levels were determined by washing the mites off all the worker bees in the alcohol samples (100-300 bees/sample) and reporting mites/100 bees. (Note: varroa counts reported here are conservative since accurate counts can only be done on bees taken from the brood area, which are not accessible in feral colonies).

Results

By the summer of 1995, as a result of many swarms during the springs of 1993-95, the feral nest sites totalled 247. By March 1996 only 12 colonies were still alive, and by March 1997, 22 were alive (Loper 1997a, see Table 1).

Morphometrics

Due to the expense and other research priorities, only a limited number of samples were analyzed by the morphometric technique. None of the 45 colonies analyzed in the fall of 1991 showed morphometric characteristics consistent with AHB; all were morphometrically EHB although samples from four colonies had forewings shorter than the standard for domestic EHB (*i.e.*, ≤ 9.01 mm). As

Table 1. Mitochondrial DNA haplotypes found in a feral honey bee population. Pinal Co. AZ 1992 - 1996

Year	Season	# Live	# Sampled	# Analyzed (n)	Sample Source	mellifera	carnical liqustica	lamarckii	scutellata
1992	Fall	159	132	132	Nests	71.2%	25.8%	2.8%	0
1993	Spring		53	53	Swarms	79.2%	18.9%	1.9%	0
1995	Spring	75	59	59	Nests	88.1%	11.9%	0	0
1995	Spring		35	29	Swarms	62.9%	14.3%	0	22.8%
1995	Fall	35	24	24	Nests	70.8%	8.3%	0	20.8%
1996	Spring	12	8	8	Nests	?*	37.5%	0	62.5%
1996	Summer	59	48	32	Nests	?*	66.7%	0	33.3%
1997	Spring	22	22		Nests				

Columns header spanning: "Colonies" over # Live, # Sampled, # Analyzed; "Haplotype (*Apis mellifera*)" over mellifera, carnical/liqustica, lamarckii, scutellata

*Tests to determine proportion of *A.m. mellifera* vs. *A.m. carnica/ligustica* were not performed.

has been found in other U.S. desert feral populations (Daly *et al.* 1991), some colonies had relatively smaller bees.

In contrast, in the 1995 data, one of six swarms (spring) and five and 29 nests (fall) showed morphometric evidence of Africanization, the probability of Africanization ranged from 0.9675 to 1.000 based on the USDAID methodology. Among the other 1995 samples analyzed morphometrically, there was a notable increase in size variation compared to the 1991 data indicating an influx of bees with smaller dimensions perhaps drifting into these colonies (Hung and Rubink 1992).

Mitochondrial DNA (mt DNA)

MtDNA data was obtained on all samples from 1992-1996 (Table 1). In 1992, 132 colonies were analyzed. Table 1 gives the number of live colonies, the number sampled, and the number analyzed *(n)*: 71.2% were maternal descendants of *A.m. mellifera*, a race of honey bee (the German honey bee). Another 25.8% were maternal descendants of *A.m. carnica/liqustica* (mtDNA methods could not distinguish between these) and 2.8% were maternal descendants of *A.m. lamarckii*, the Egyptian honey bee. These latter four colonies were all located within 700 m of each other. To our knowledge *A.m. lamarckii* has not been introduced into the U.S. for over 100 years (but see Stephens 1995). No *A.m. scutellata* haplotypes were found in any of the 1992 and 1993 samples (185 colonies).

In spring 1993 samples, there was no significant change in the proportions of haplotypes from those of 1992 (Pearsons X^2=1.27, df=2, P>0.53). A resampling of all live, accessible feral colonies in the study area was done on March 25-26, 1995. By this time, tracheal mite parasitism had reduced the live colony population to 72 (Loper 1995a). Mt DNA analyses revealed a continued, and now significant (X^2=5.9, df=2, P=0.05) shift in the proportions of maternal haplotypes: of the 59 colonies analyzed, the proportion of *A.m. mellifera* was 0.881, and the rest were *A.m.carnica liqustica* (Table 1). There were no AHB haplotypes (either *A.m. lamarckii* or *A.m. scutellata*) in the spring 1995 established colonies. However, the mt DNA of 35 swarms caught in the spring of 1995 showed (Table 1) that eight (22.8%) were AHB (*A.m. scutellata*). These were the first AHB swarms documented in the study area.

In the fall of 1995, 17 colonies had the mt DNA haplotypes of *A.m.mellifera*, two had *A.m. carnica/*

ligustica, and five had *A.m. scutellata* (Table 1). These five colonies are the first documented AHB colonies established in the rock nest sites in the study area.

There was a further die-off of these colonies during the winter of 1995-96 reducing the population to 12 colonies in March, 1996. Eight of these 12 colonies were accessible, sampled and analyzed; five had AHB mt DNA haplotypes. The population increased during the spring of 1996 to 59 colonies: 33% of those sampled (N=30) had AHB mt DNA.

Allozymes

The shift in allozyme frequencies between 1992 and the summer of 1996 is shown in Table 2. The Mdh allelic frequencies of 1992, 1994 and early 1995 are all very similar, indicating a genetically stable population. The first tentative indication of Africanization was detected by a low level (0.10) of Hk[83] in a sample of March 29, 1994 (the result of one bee of the 24 analyzed from that colony having both Hk[100] and Hk[83] allozymes). However, no other feral colonies sampled in the fall of 1994 or the spring of 1995 had Hk[83]. Allelic evidence of AHB immigration into the study area was first seen in the swarms of 1995, consistent with the mt DNA data. The eight swarms with AHB mt DNA had significantly higher proportions of Mdh[100] (=0.60) vs. the EHB (mt DNA) swarms (=0.17)(Table 2). Even though these 35 swarms were removed from the area (or died from varroa mites), the fall 1995 sampling of feral colonies showed an increase in Mdh[100] and Hk[83] associated with Africanization in both the EHB and AHB mt DNA haplotypes (Table 2).

Mites

On Dec. 12 and 13, 1996, 25 of the 28 live colonies were sampled for tracheal and varroa mites. Tracheal mites were present in 88% of the colonies (range 3.3 to 36.7% infested) and varroa mites were found in 72% of the colonies (max.=29% infestation).

Discussion

Before 1995, this population of feral honey bees was relatively genetically stable, providing a baseline for the evaluation of the effects of mite parasitism and Africanization. Surprisingly, the tracheal mite parasitism (initiated as early as 1991, but heaviest in 1993-94 Loper 1995a) related to an increase in the proportion of *A.m. mellifera* mt DNA

Table 2. Shift in genetics (allozymes and mitochondrial DNA) of feral honey bee colonies—Pinal Co. AZ 1992-1996

Date	Season	# of colonies analyzed(n)	(n) as a % of live colonies	Source of sample	Mdh 100	80	65	Hk 100	83	mt DNA
1992	Fall	132	83.0	Nest	.20	.31	.49			EHB
1994	Spring	12	10.5	Nest	.26	.19	.55	.98	.02*	EHB
1994	Fall	7	5.3	Nest	.33	.16	.51	1.00		EHB
1995	Spring	14	18.7	Nest	.24	.27	.49	.99	.01	EHB
1995	Fall	9	13.2	Nest	.45	.30	.25	.96	.04	EHB
1995	Fall	5	7.4	Nest	.77	.14	.09	.93	.07	AHB
1995	Fall TOTAL	14	20.6	"	.57	.24	.19	.95	.05	EHB & AHB
1995	Spring	27	77.1	Swarms	.17	.23	.60	-	-	EHB
1995	Spring	8	22.9	Swarms	.60	.12	.28	-	-	AHB
1995	Spring TOTAL	35		"	.27	.21	.52	-	-	EHB & AHB
1996	Summer	19	32.8	Nest	.43	.25	.32	1.00	0.0	EHB
1996	Summer	30	51.7	Nest	.61	.18	.21	.95	0.05	AHB**
1996	Summer TOTAL	49		"	.54	.20	.26	.97	0.03	

* 1 colony had the first indication of hybridization with AHB by having .10 Hk[83], the European queen having mated with at least 1 AHB drone.
** Called AHB based on MDH 100>0.67 and/or Hk[83] >0.0, mt DNA not available.

haplotypes relative to *A.m. carnica/ligustica*. However, the biggest effect of the tracheal mites was the reduction of the EHB colony population size.

The preferential infestation and reproduction of *Varroa* in drone brood (Schulz 1984) drastically reduces drone production: whereas literally thousands of drones could be (and were) trapped using aerial net traps in the study area before mite parasitism, I (GML) was unable to attract or trap more than 3-4 drones in 1996. However, on June 2, 1997, 131 drones were caught in the study area in one hour (Loper unpublished). Thus, mite parasitism almost eliminated the EHB genetic "buffer" that existed just before Africanization began in the study area.

The numbers of live colonies that overwintered were lowest in March 1996, but in both 1996 and 1997, we documented (genetically) that some of these colonies were AHB and were the result of early season swarming. This is consistent with AHB behaviors including shorter development time, faster colony development and increased swarming and absconding (Winston *et al.* 1979; Loper 1995b, Rubink *et al.* 1996).

None of the three genetic tests used in this study are adequate by themselves to determine the levels of genetic shifts occurring in a population. Each method (morphometrics, mtDNA, Hk allozymes) can detect Africanization in samples that can be missed by each of the other methods (Loper 1997b). However, when taken together, especially the use of mtDNA and Hk analyses, the introgression of AHB genes can be determined (Smith *et al.* 1989; Taylor *et al.* 1991; Rubink *et al.* 1995). Even before Africanization, morphometric analyses documented that the population in this study had colonies that were different, *i.e.*, "feral", in size characteristics compared to the domestic honey bee of general commercial use.

Continued seasonal sampling of these feral colonies will be needed to determine if the population once again reaches genetic stability and to determine the degree of Africanization of the feral population in southern Arizona.

Literature

Burgett, M. 1996. PNW Colony Mortality 1996. *National Honey Market News*, p. 10. August 9, 1996.

Daly, H.V., K. Hoelmer, and P. Gambino. 1991. Clinal geographic variation in feral honey bees in California, USA. *Apidologie* 22:591-609.

Delfinado-Baker, M. 1984. *Acarapis woodi* in the United States. *Am. Bee J.*124:805-806.

Del Lama, M.A., R.A. Figueiredo, A.E.E. Soares and S.N. Del Lama. 1988. Hexokinase polymorphism in *Apis mellifera* and its use for Africanized honey-bee identification. *Rev. Bras. Genet.* 11:287297.

Hung, A., C.F. and W.L. Rubink. 1992. Biochemical evidence of nonprogeny workers in feral Africanized honey bee (*Apis mellifera* L.) colonies (Hymenoptera: Apidae. *BeeScience* 2:33-36.

Kraus, B. and R.E. Page, Jr. 1995. Effect of *Varroa jacobsoni* (Megostigmata: Varroidae) on feral *Apis mellifera* (Hymenoptera: Apidae) in California. *Environ. Entomol.* 24:1473-1480.

Loper, G.M. 1995a. A documented loss of feral bees due to mite infestations in S. Arizona. *Am. Bee J.* 135:823-824.

Loper, G.M. 1995b. Some attributes of the Africanized honey bees in Southern Arizona: wing length, hygienic behavior, worker emergence time and brood nest temperature. *Am. Bee J.* 135:828 (Abstract).

Loper, G.M. 1996a. Feral colonies and tracheal mites. *Bee Culture* 124:27.

Loper, G.M. 1997a. Overwinter losses of feral honey bee colonies in Southern Arizona, 1992-1997. *Am. Bee J.* 137:446.

Loper, G.M. 1997b. Genetic evidence of the Africanization of feral colonies in S. Arizona between 1993 and 1995. *Am. Bee J.* 137: 669-671.

Rinderer, T.E., S.M. Buco, W.L. Rubink, H.V. Daly, J.A. Stelzer, R.M. Riggio, and F.C. Baptista. 1993. Morphometric identification of Africanized and European honey bees using large reference populations. *Apidologie* 24:569-585.

Rubink, W.L., P. Luévano-Martinez, W.T. Wilson, and A.M. Collins. 1995. Comparative Africanization rates in feral honey bee populations at three latitudes in Northeastern Mexico and Southern Texas. (Abstract) *Am.*

Bee J. 135:829.

Rubink, W.L., P. Luévano-Martinez, E.A. Sugden, W.T. Wilson, and A.M. Collins. 1996. Subtropical *Apis mellifera* (Hymenoptera: Apidae) swarming dynamics and Africanization rates in Northeastern Mexico and Southern Texas. *Ann. Entomol. Soc. Am.* 89:243-251.

Schiff, N.M., W.S. Sheppard, G.M. Loper, and H. Shimanuki. 1994. Genetic diversity of feral honey bee (Hymenoptera: Apidae) populations in the Southern United States. *Ann. Entomol. Soc. Am.* 87:842-848.

Schmidt, J.O., S.C. Thoenes, and R. Hurley. 1989. Swarm traps. *Am. Bee J.* 129:468-471.

Schulz, A.E. 1984. Reproduktion und Populationsentwicklung derparasitischen Milbe *Varroa jacobsoni* Oud. in Abhangigkeit vom Brutzyclus ihres Wirtes *Apis mellifera* L. *Apidologie* 15:401-420.

Sheppard, W.S. 1988. Comparative study of enzyme polymorphism in United States and European honey bee (Hymenoptera: Apidae) populations. *Ann. Entomol. Soc. Am.* 81:886-889.

Sheppard, W.S., A.E.E. Soares, D. DeJong, H. Shimanuki. 1991. Hybrid status of honey be populations near the historic origin of Africanization in Brazil. *Apidologie* 22:643-652.

Smith, D.R., O.R. Taylor, Jr., G. Loper, and W. Rubink. 1989. Inferring the matrilineal descent of feral American bees using mitochondrial DNA and allozymes. *Am. Bee J.* 129:823.

Spivak, M., T. Ranker, O. Taylor, Jr., W. Taylor, and L. Davis. 1988. Discrimination of Africanized honey bees using behavior, cell size, morphometrics, and a newly discovered isozyme polymorphism. pp. 313-324 *In* G.R. Needham, R.E. Page, Jr., M. Delfinado-Baker, and C.E. Bowman [eds.], Africanized honey bees and bee mites. Horwood, Chichester, U.K.

Taylor, O.R. Jr., A. Delgado, and F. Brizuela. 1991. Rapid loss of European traits from feral neotropical African honey bee populations (Abstract) *Am. Bee J.* 131:783-784.

Stephens, C.A. 1995. The old squire's Egyptians. *Am. Bee J.* 135:809-810.

United States Department of Agriculture, National Agricultural Statistics Service, Agricultural Statistics Board. 1992 Statistical Bulletin #912.

Winston, M.L., G.W. Otis, O.R. Taylor, Jr. 1979. Absconding behaviour of the Africanized honeybee in South America. *J. Apic. Res.* 85-94.

Genetic and physiological studies of African and European honey bee hybridization: past, present, and into the 21st century

H. GLENN HALL
DEPARTMENT OF ENTOMOLOGY AND NEMATOLOGY
UNIVERSITY OF FLORIDA
GAINESVILLE, FL 32611 U.S.A.

Phone: 352-392-1901
Fax: 352-392-0190
email: hgh@gnv.ifas.ufl.edu

Abstract

Population studies with DNA markers have revealed that gene flow between African and European honey bees (*Apis mellifera* L.) in the neotropics has been asymmetric in favor of African bees. Evidence from physiological studies indicates that hybrid deficiencies may be involved. The findings from these studies are reviewed and discussed, and arguments against interpretations of the findings are addressed. Directions in which this research may go into the 21st century are suggested.

Introduction

For the past twelve years, studies by me and my research group have involved the use of DNA to distinguish African and European honey bees. My original purpose was to find a reliable means of identification that would be needed for regulatory control of the African bee. Other methods of identification were inadequate (reviewed by Daly 1988, 1991). Shortly after I began to find distinguishing markers and to test samples of bees from the neotropics, the markers began to reveal interesting information about the processes involved in the expansion of the African bee population. The early findings suggested that the invading African bees and the resident European bees were not forming a "hybrid swarm" population as had been assumed (Rinderer 1986). Introgression was asymmetric: far greater gene flow from feral African colonies into managed European colonies, than in the opposite direction. After African bees became established in an area, a mostly African population was the result. European bees disappeared, along with hybrids carrying European genes. However, more recent DNA evidence has revealed a substantial European contribution to the neotropical African population. (These points will be discussed in more detail below). The DNA findings have raised questions about the genetic relationships among honey bee subspecies and have encouraged further research to learn more about African and European honey bee hybridization.

DNA markers

In our studies, both mitochondrial DNA (mtDNA) and nuclear DNA are used as sources of markers. MtDNA is a short circular piece found in the energy-producing organelles of the cell, the mitochondria. In most higher animals, including the honey bee, mtDNA is inherited only from the mother (Brown 1985, Avise 1986), and characteristic differences serve as markers of maternal lineages. The DNA in the nucleus of the honey bee is more than ten thousand times longer than mtDNA (Jordon and Brosemer 1974, Crain *et al.* 1976, Smith and Brown 1990) and carries most of the genetic information. If there are two parents, nuclear DNA is inherited from both.

We have primarily used DNA restriction fragment length polymorphisms (RFLPs) as markers. Nuclear DNA RFLPs were detected in Southern blots (Southern 1975), using anonymous fragments of cloned honey bee DNA as radioactive probes (Hall 1986, 1990, 1991, 1992a, McMichael and Hall 1996). The Southern blot approach was very effective but was time-consuming and difficult for analyzing many samples. To make DNA testing easier, we have employed the polymerase chain reaction (PCR)(Saiki *et al.* 1988). DNA regions carrying RFLPs are amplified, rather than being detected with probes. We first used this approach, now referred to as PCR-RFLP, to facilitate identification of African and European honey bee mtDNA (Hall and Smith 1991). Since then, we have converted most of our nuclear DNA markers, found initially with cloned probes, so that they can be analyzed using the PCR (Hall 1998, Hall *et al.* 1999). More recently, we have found new RFLP markers in DNA that is first amplified (approach from Karl and Avise 1993) but with modified procedures to amplify long regions (Barnes 1994, Cheng *et al.* 1994), so that the chances of finding useful markers are greatly increased (Suazo and Hall 1999).

Most regions of DNA in which we have found RFLPs are present as single copies and can be considered genetic loci. Thus, the different polymorphic forms from a region are considered alleles. Some loci have only two alleles, representing a single restriction site difference. Other loci have many alleles, as many as 85, representing combinations of several site differences, from one or more enzymes, plus length differences (insertions, deletions, and tandem duplications). Typically, at the different loci, we have found some alleles specific to either African or European bees, with the other alleles common to both. We have not found any allele that alone is "diagnostic", that is, present at a 99% frequency in one bee type and absent in the other (accepted standard, Ayala and Powell 1972). However, for some loci, African or European-specific alleles are present collectively at frequencies above 99%. At the locus with 85 alleles mentioned above, almost all the alleles are either west European, east European or African (McMichael and Hall 1996).

Other types of markers based on the PCR are commonly used: RAPDs (random amplified polymorphic DNA; Williams *et al.* 1990, reviewed in Hadrys *et al.* 1992) and microsatellites (Tautz 1989). We have found a few RAPD markers that distinguish African and European bees (Suazo, McTiernan, Hall 1998). We prefer PCR-RFLP markers which are far more reproducible and less subject to artifacts common to RAPDs (Halldén *et al.* 1996, and references therein). Most RAPD markers are dominant, whereas PCR-RFLP markers are co-dominant, allowing heterozygotes to be identified, and, thus, are more useful for population studies. With microsatellites, more complicated electrophoretic procedures are needed to distinguish the small differences in fragment sizes, but, once separated, the single fragments representing the alleles are easier to identify in heterozygotes, than some PCR-RFLP alleles that have several fragments. Microsatellite alleles may be evolutionarily less stable (Amos *et al.* 1996).

For a more detailed discussion of DNA markers, see Hall (1991)

Processes involved in New World African honey bee spread revealed with DNA markers

Eight years ago, Deborah Smith, now at the University of Kansas, and I independently found African mtDNA in virtually all feral swarms in the neotropics, including those comprising the expanding front of the African population (Hall and Muralidharan 1989, Smith *et al.* 1989, Hall and Smith 1991). These results revealed that neotropical African bees had spread as swarms that were continuous African maternal lineages extending back to the bees introduced from Africa. This finding contradicted previous views that the neotropical population consisted primarily of Africanized progeny of African drones mated to European queens in apiaries. There is no debate that these matings occur, resulting in considerable paternal introgression into European apiaries. The change in the behavior of managed bees, most notably in their stinging, is dramatic. With DNA markers, evidence was obtained consistent with such African paternal gene flow into apiaries (Hall 1990). However, studies of managed apiaries rather than of feral colonies have given an exaggerated impression of the importance of such matings (Rinderer *et al.* 1985, Rinderer 1986). Despite repeated backcrossing to feral African drones, swarms from these Africanized apiaries do not become a significant part of the expanding feral African population. After such backcrossing, Africanized European matrilines would have mostly African nuclear DNA but would still have European mtDNA, but very little European mtDNA is found in the feral population after African bees become established. Actively maintained European colonies probably account for the few European swarms which tend to be localized (Hall and McMichael 1992, Lobo 1995, Quezada-Euan and Hinsull 1995). Unless actively maintained, European matrilines in apiaries eventually disappear, probably through attrition. African matrilines enter apiaries as feral swarms invading colonies, inhabiting empty equipment, or being hived by beekeepers to replace lost colonies (Michener 1975, Taylor 1985a,b). Instead of being the main contributor to the neotropical African population, paternal Africanization of European colonies is a dead-end process.

DNA and other results have suggested that some hybridization of the invading feral African bees occurs as they first encounter populations of managed European bees, but the hybrids are apparently lost as the African population becomes established. In feral African swarms near the front, low levels of east European nuclear DNA markers were found, but in populations further behind the front, east European markers were virtually absent (Hall 1990). These results coincide with temporal changes towards a more African morphology seen in Panama (Boreham and Roubik 1987) and in allozyme frequencies seen in Mexico (Taylor et al. 1991).

More recent DNA evidence indicates that the gene pool of neotropical African bees has a west European contribution (e.g., from A. m. mellifera). The earlier DNA studies used markers specific to east European bees (e.g., A.m. ligustica)(Hall 1990). Consistent with the earlier findings, east European markers at another locus were absent in feral neotropical African colonies. However, west European markers at the same locus were found at substantial levels, about 20% to 30% (Hall and McMichael 1998). There was not a consistent increase in frequency of these markers in a northward direction, as a result of the African bees encountering additional European populations. We believe that the markers may have entered the African population shortly after African bees were introduced. If New World African bees descended primarily from virgin queens distributed to beekeepers (Rinderer et al. 1993), considerable hybridization might have occurred with the resident bees, which, at that time, in that area of Brazil, were predominantly west European (Kent 1988). Subse-

quent selection could have eliminated most European genes but may have left genes and/or markers that were neutral. Hybridization of neotropical African bees with European bees has been suggested from morphometric scores (Buco et al. 1987). Hybridization specifically with west European bees has been suggested from allozyme frequencies (Lobo et al. 1989, Lobo and Krieger 1992).

For past, more detailed, reviews on the DNA findings, see Hall (1991, 1992b)

Mechanisms that may limit hybrid survival and favor the African genotype

Probably, strong adaptation to the tropical environment is responsible for African bee dominance, and poor adaptation is largely responsible for the loss of European bees and their hybrids. I have suggested that hybrid survival would be further impaired, if incompatibilities between African and European bees existed that led to hybrid breakdown or dysfunction (Hall and Muralidharan 1989, Hall 1991, 1992b). The failure of European mtDNA to persist in the feral African population indicates that it is not neutral and, if there were incompatibilities, differences in metabolism might be involved. African bees have a shorter developmental period and a shorter lifespan (reviewed in Winston et al. 1983). African swarms disperse far greater distances (Lindauer 1955, Fletcher 1978, Taylor 1977). African and European bees exhibit differences in thermoregulation (Southwick et al. 1990) and significant differences in allele frequencies of metabolic enzymes hexokinase and malate dehydrogenase (reviewed in Daly 1991). All these characteristics implicate differences in their metabolism.

To test the hypothesis that hybrids may have metabolic deficiencies, we produced a number of first generation and backcross hybrids through artificial insemination at the Escuela Agrícola Panamericana in Honduras. In collaboration with Jon Harrison at Arizona State University, we determined metabolic rates based on CO_2 production of departing forging workers during agitated flight in a small chamber (Harrison and Hall 1993). The African bees weighed less than the European bees, and the hybrids were intermediate between the two. The African bees had higher mass-specific metabolic capacities than the Europeans. Hybrids had metabolic capacities that were not intermediate but, rather, were equal to or lower than the Europeans. Smaller insects use more energy to sustain hovering flight, but the differences in metabolic capacities could not be explained by the differences in size alone. When the sizes were taken into account, the actual metabolic rates of African bees, in comparison to the European bees, were higher than expected and the actual rates of the hybrids were lower than expected. These results suggest that African bees are superior to European bees in flight performance, and hybrids are inferior. Because first generation hybrids from reciprocal crosses both had a reduced metabolic rate, a European maternal component could not be completely responsible. However, when the data were pooled from first generation and backcross hybrids according to matrilines, the metabolic rates of hybrids from European matrilines were significantly lower than the rates of hybrids from African matrilines. Higher metabolic capacity can enhance several aspects of flight performance such as speed, maneuverability, and the ability to carry heavy loads. These traits have yet to be evaluated directly. Because flight is a fundamental aspect of bees' existence, it is logical that inferior flight performance would be a selective disadvantage in hybrids, which may be a factor that has helped preserve the African genotype in the neotropics.

Other factors could favor the African genotype. For example, daughter queens of African paternity might be selected through faster developmental times (Winston et al. 1983, Taylor 1988) and/or neopotism (Page et al, 1989), whereby their supersister workers begin to rear them earlier. If so, earlier emerging queens of African paternity would have a major advantage, because they would kill younger queens of European paternity still in their cells (Taylor 1988). This would be an effective mechanism of excluding European alleles from the reproductive population that might be found in the workers (Hall 1991).

Arguments against interpretations of the genetic and physiological findings

It has been argued that the greater fitness of African bees in a tropical environment, resulting in overwhelming numbers, alone could account for the paucity of European mtDNA in the neotropics, and that some type of hybrid dysfunction does not have to be invoked (Page 1989). Along similar lines, it has been argued that the lack of European mtDNA in feral swarms is because European colonies swarm less frequently than African colonies. As

mentioned above, I believe that selection in a tropical environment is primarily responsible for the dominance of African bees and the loss of European bees and their hybrids. However, among the African traits that European matrilines in apiaries should acquire, after repeated backcrossing with African drones, are the tendency to generate more swarms and the ability to survive in tropical environments. Such Africanized swarms should have accumulated in the feral population over the distances and over the time that the African bees have spread in the New World. Because they have not, I therefore speculated that European mtDNA was not neutral or might be involved in some type of hybrid breakdown.

It has also been argued that the limited hybridization of feral African bees, observed with DNA markers, does not reflect hybrid deficiencies (poor adaptation to the tropics or dysfunction) but, rather, reflects a lack of European colonies in the areas in which the samples were collected (Rinderer et al. 1991, 1993). Some apparent hybridization (based on morphometrics and mtDNA) was observed in the Yucatan peninsula of Mexico, where there has been a high concentration of European colonies (Rinderer et al. 1991). There is little disagreement that where there is a source of African and European bees there will be hybridization. As mentioned above, some hybridization, based on nuclear DNA and mtDNA, was seen also near the expanding front of the African population (i.e., with east European bees, above the preexisting level with west European bees)(Hall 1990). However, the key point is whether such hybrids persist. The evidence described above suggests that they do not. In the tropics, where African bees have a greater fitness, the European bees are eventually replaced, thus eliminating the source for forming hybrids. The larger the European population, the longer time the replacement will take (Rinderer et al. 1993, Quezada-Euan and Hinsull 1995). A source of European colonies could be actively maintained, but, in the tropics, this effort would face strong selective pressure and would have to be continuous. Because African bees disperse long distances (Taylor 1977), higher levels of Europeanization in some areas would indicate that the effect is local and is not sustained upon dispersal. In temperate regions, where European bees have a greater fitness, they would remain as a source that could form hybrids. Near the subtropical-temperate boundaries, African and European bees and their hybrids may be equally adapted, and hybrids may persist in zones between the two types of bees (Taylor 1977, Taylor and Spivak 1984, Lobo *et al.* 1989, Sheppard *et al.* 1991). Even if factors other than the environment select against hybrids, they could continue to be formed from parental African and European bees moving into the hybrid zones.

Although, the failure of Africanized European matrilines to become part of the feral population is what led me to suggest that hybrid breakdown might be involved, such dysfunction could occur also in hybrids from African matrilines and tend to eliminate them from the feral population. The metabolic deficiencies we observed in hybrids appears to be the type of dysfunction that had been hypothesized.

Into the 21st Century:
Honey bee population studies

More robust population studies will be possible with DNA markers that can be more easily used, such as PCR-RFLPs. More samples can be tested with more markers, thereby providing greater resolution of introgression that may take place between African and European bees. More thorough studies of hybridization in the neotropics are needed to substantiate the previous findings and to provide further details. Levels of hybridization at different locations and over time at the same locations need to be investigated. African and European matrilines would continue to be evaluated separately, which, as before, would likely have different outcomes. Multiple alleles at different nuclear loci will enable parental analyses. Differences between maternal and paternal genotypes might point to selective pressures that act preferentially against workers, queens or drones.

African bees have now moved into nearly all continental neotropical areas that they are likely to occupy. Hybridization may still be studied in areas of the neotropics where large resident European populations are still in the process of being replaced and/or where European bees are being maintained. However, opportunities have largely passed for real time studies of the expanding African front in tropical areas and their initial encounters with European populations. Orley Taylor, University of Kansas, and William Rubink, USDA-ARS, Weslaco, have had the foresight to collect thousands of swarms in northern Mexico over a period of several years, when African bees were becoming established. From

such samples remain the chance to resolve some questions that still remain about the African bee takeover.

Some of the most valuable population studies in the future could come from the predicted transition or hybrid zones (Taylor and Spivak 1984). The genetic structure of the zones may point to the nature of selective factors (Barton and Hewitt 1985, 1989, Harrison 1990). In the subtropical-temperate boundary, the differential effects of tropical or temperate environmental selection in favor of the African or European genotype should be reduced or eliminated. The effects of other factors on hybridization, if they exist, may be distinguished. If only ecological factors are involved, hybrids should exist in equilibrium with the parental types within a broad zone. Genes responsible for adaptive traits, along with closely linked "hitchhiking" markers, may form reciprocal clines. A hybrid population would serve as a matrix for introgression, and some neutral markers should introgress readily beyond the zone. The introgressive behavior of mtDNA and nuclear DNA markers might reflect whether they are neutral or non-neutral, or whether they are linked to, or have functional associations with, non-neutral genes. If factors other than the environment exist, which reduce hybrid survival, such as hybrid dysfunction, African and European genotypes would tend to remain separate. Genotype disequilibrium, with a deficiency of heterozygotes, and gametic phase disequilibrium would persist within the zone. Introgression beyond the zone of even neutral traits would be impeded.

Adding further to the devastating damage being caused by the *Varroa* mite, the opportunity to study the putative African-European honey bee hybrid zone in the US may be greatly limited and eventually lost. In addition to the samples collected in northern Mexico, Orley Taylor and William Rubink have been collecting feral swarms samples from southern Texas. Some of the information described above should still be accessible through such samples.

Into the 21st Century:
Studies of individual and colony level processes affecting hybridization

A number of processes at the individual or colony level could influence the amount of hybridization between African and European bees and favor the African genotype. Some assortative mating has been detected using visual markers (Kerr and Bueno 1970) and more recently using allozymes (Orley Taylor, personal communication). Additional studies using a collection of African and European-specific PCR-RFLP markers could provide more substantial evidence. Parental analyses using the loci with multiple PCR-RFLP alleles should be valuable in studies attempting to determine if there is preferential selection of queens of African paternity.

The fundamental difference between African and European bees may prove to be in their metabolism, perhaps a consequence of adaptation to tropical and temperate environments. Alleles of some enzymes, that have been used primarily as genetic markers, namely malate dehydrogenase, appear not to be neutral. Alleles exist in clines in different continents (Nielsen *et al.* 1994, Oldroyd *et al.* 1995), influence metabolic rates (Harrison *et al.* 1996) and have unequal thermostability (Cornuet *et al.* 1995). Many enzymes are heteromeric, that is composed of different subunits. Enzymes composed of subunits from different subspecies may not have optimal activity. Differences between African and European matrilines might be a consequence of mitochondrial enzymes composed of both mitochondrial and nuclear DNA encoded subunits. The apparent deficiencies in hybrid metabolic capacities would be consistent with a disruption of such enzyme complexes, where European mitochondrial-encoded subunits in combination with African nuclear subunits are more detrimental than the reciprocal combination. Besides flight capacity, a number of other critical aspects of hybrid bee biology could be affected by metabolic deficiencies.

Concluding remarks

Less than a decade ago, honey bee genetic studies were limited because few genetic markers were available. Thus, some of our views about African honey bees and their interaction with European bees arose from assumptions. Many other questions about bee genetics were not even approachable. Constantly improving DNA technology has been providing us with a wealth of markers compared to what we had previously. Within a short time, RFLPs have increased our understanding of the African bee. RFLPs, RAPDs, and microsatellites have also helped identify other population structures (Smith *et al.* 1991), resolve phylogenetic relationships among the bee races (Smith 1991, Garnery *et al.* 1992, Estoup *et al.* 1995), study in-

tra-colony relationships (Fondrk *et al.* 1993, Estoup *et al.*, 1994, Oldroyd *et al.* 1994), construct genetic maps (Hunt and Page 1995), and localize genes responsible for sex determination and behavior (Hunt and Page 1994, Hunt *et al.* 1995). By integrating the "new" genetics with physiological, behavioral, neurological, ecological and evolutionary studies, advances in bee biology in the first part of the 21st century promise to be exciting and impressive.

Acknowledgments

My work has been supported by grants from the USDA National Research Initiative Competitive Grants Program. Additional funds have been obtained from the US Agency for International Development, Florida-Costa Rica Institute (FLORICA), the University of Florida Interdisciplinary Center for Biotechnology Research (ICBR), and the Florida State Beekeepers Association.

I am grateful to the many people who have helped provide honey bee samples over the years for my studies. United States: Steve Taber, Norman Gary, Orley Taylor, Marla Spivak, Rick Hellmich, Jose Villa, Anita Collins, Tom Rinderer, Deborah Smith, Gerald Loper, Gordon Waller, Joe Martin, Eric Erickson, William Rubink, Willem Van der Put, Tom Sanford. Mexico: Felipe Brizuela. Honduras: Alonso Suazo. Costa Rica: Henry Arce. France: Bernard Vaissiere, Jean-Marie Cornuet. South Africa: Robin Crewe, George Pretorius, and many beekeepers.

Keywords

African honey bees, European honey bees, hybridization, gene flow, introgression, population genetics, genetic markers, mitochondrial DNA, nuclear DNA, DNA polymorphism, RFLP, polymerase chain reaction, PCR, metabolism, hybrid dysfunction, isolating mechanisms

References cited

Amos, W., S. J. Sawcer, R. W. Feakes, and D. C. Rubinsztein. 1996. Microsatellites show mutational bias and heterozygote instability. *Nature Genet.* 13: 390-391.

Avise, J. C. 1986. Mitochondrial DNA and the evolutionary genetics of higher animals. *Phil. Trans. R. Soc. London B* 312: 325-342.

Ayala, F., and J. Powell. 1972. Allozymes as diagnostic characters of sibling species of *Drosophila. Proc. Natl. Acad. Sci. U.S.A.* 69: 1094-1096.

Barnes, W. M. 1994. PCR amplification of up to 35-kb DNA with high fidelity and high yield from lambda bacteriophage templates. *Proc. Natl. Acad. Sci. U.S.A*

91: 2216-2220.

Barton, N. H., and G. M. Hewitt. 1985. Analysis of hybrid zones. *Ann. Rev. Ecol. Syst.* 16: 113-148.

Barton, N. H., and G. M. Hewitt. 1989. Adaption, speciation and hybrid zones. *Nature* 341: 497-503.

Boreham, M. M., and D. W. Roubik. 1987. Population change and control of Africanized honey bees (Hymenoptera: Apidae) in the Panama canal area. *Bull. Ent. Soc. Amer.* 33: 34-38.

Brown, W. M. 1985. The mitochondrial genome of animals. pp. 95-130. In: *Molecular Evolutionary Genetics.* R. J. MacIntyre [ed.]. Plenum Press, New York.

Buco, S. M., T. E. Rinderer, H. A. Sylvester, A. M Collins, V. A. Lancaster, and R. M. Crewe. 1987. Morphometric differences between South American Africanized and South African (*Apis mellifera scutellata*) honey bees. *Apidologie* 18: 217-222.

Cheng, S., C. Fockler, W. M. Barnes, and R. Higuchi. 1994. Effective amplification of long targets from cloned inserts and human genomic DNA. *Proc. Natl. Acad. Sci. U.S.A* 91: 5695-5699.

Cornuet, J -M., B. P. Oldroyd, and R. H. Crozier. 1995. Unequal thermostability of allelic forms of malate dehydrogenase in honey bees. *J. Apicultural Res.* 34: 45-47.

Crain, W. R., E. H. Davidson, and R. J. Britten. 1976. Contrasting patterns of DNA sequence arrangement in *Apis mellifera* (honeybee) and *Musca domestica* (housefly). *Chromosoma* 59: 1-12.

Daly, H. V. 1988. Overview of the identification of Africanized honey bees. pp. 245-249. In: *Africanized Honey Bees and Bee Mites.* G. R. Needham, R. E. Page, M. Delfinado-Baker, C. E. Bowman [eds.]. Ellis Horwood Limited, Chichester.

Daly, H. V. 1991. Systematics and identification of Africanized honey bees. pp. 13-44. In: *The "African" Honey Bee.* M. Spivak, M. Breed, D. J. C. Fletcher [eds.]. Westview Press, Boulder.

Estoup, A., M. Solignac, and J. -M. Cornuet. 1994. Precise assessment of the number of patrilines and of genetic relatedness in honeybee colonies. *Proc. R. Soc. Lond.* B 258: 1-7.

Estoup, A., L. Garnery, M. Solignac, and J. -M. Cornuet. 1995. Microsatellite variation in honey bee (*Apis mellifera* L.) populations: Hierarchical genetic structure and test of the infinite allele and stepwise mutation models. *Genetics* 140: 679-695.

Fletcher, D. J. C. 1978. The African bee, *Apis mellifera adansonii*, in Africa. *Ann. Rev. Entomol.* 23: 151-171.

Fondrk, M. K., R. E. Page, Jr., and G. J. Hunt. 1993. Paternity analysis of worker honeybees using random amplified polymorphic DNA. *Naturwissenschaften* 80: 226-231.

Garnery, L., J. -M. Cornuet, and M. Solignac. 1992. Evolutionary history of the honey bee *Apis mellifera* inferred from mitochondrial DNA analysis. *Molec. Ecol.* 1: 145-154.

Hadrys, H., M. Balick, and B. Schierwater. 1992. Applications of random amplified polymorphic DNA (RAPD) in molecular ecology. *Molec. Ecol.* 1: 55-63.

Hall, H. G. 1986. DNA differences found between Africanized and European honeybees. *Proc. Natl. Acad. Sci. U.S.A.* 83: 4874-4877.

Hall, H. G. 1990. Parental analysis of introgressive hybridization between African and European honeybees using nuclear DNA RFLPs. *Genetics* 125: 611-621.

Hall, H. G. 1991. Genetic Characterization of honey bees through DNA analysis. pp. 45-73. In: *The "African" Honey Bee.* M. Spivak, M. Breed, D. J. C. Fletcher [eds.]. Westview Press, Boulder.

Hall, H. G. 1992a. Further characterization of nuclear DNA RFLP markers that distinguish African and European honeybees. *Arch. Insect Biochem. Physiol.* 19: 163-175.

Hall, H. G. 1992b. Processes of New World African honeybee spread revealed by DNA studies. *Florida Ent.* 75: 51-59.

Hall, H. G. 1998. PCR amplification of a locus with RFLP alleles specific to African honey bees. *Biochemical Genetics.* 36: 351-361.

Hall, H. G., and K. Muralidharan. 1989. Evidence from mitochondrial DNA that African honey bees spread as continuous maternal lineages. *Nature* 339: 211-213.

Hall, H. G., and D. R. Smith. 1991. Distinguishing African and European honey bee matrilines with amplified mitochondrial DNA. *Proc. Natl. Acad. Sci. U.S.A.* 88: 4874-4877.

Hall, H. G., and M. McMichael. 1998. The gene pool of neotropical African honey bees has a substantial contribution from west European races but not from east European races: evidence from RFLP allele frequencies at a polymorphic locus. In preparation.

Hall, H. G., A. Suazo, and R. McTiernan. 1998. A highly polymorphic anonymous DNA locus of the honey bee useful for parental analyses. In preparation.

Halldén, C., M. Hansen, N. -O. Nilsson, A. Hjerdin, and T. Säll. 1996. Competition as a source of errors in RAPD analysis. *Theor. Appl. Genet.* 93: 1185-1192.

Harrison, J. F., and H. G. Hall. 1993. African-European honeybee hybrids have low non-intermediate metabolic capacities. *Nature* 363: 258-260.

Harrison, J. F., D. I. Nielsen, and R. E. Page, Jr. 1996. Malate dehydrogenase phenotype, temperature and colony effects on flight metabolic rate in the honeybee, *Apis mellifera. Funct. Ecology* 10: 81-88.

Harrison, R. G. 1990. Hybrid zones: windows on the evolutionary process. pp. 69-128. In: *Oxford Surveys in Evolutionary Biology.* Volume 7. D. J. Futuyma and J. Antonovics [eds.]. Oxford University Press, Oxford.

Hunt, G. J., and R. E. Page, Jr. 1994. Linkage analysis of sex determination in the honey bee (*Apis mellifera* L.). *Mol. Gen. Genet.* 244: 512-518.

Hunt, G. J., and R. E. Page, Jr. 1995. Linkage map of the honey bee, *Apis mellifera*, based on RAPD markers. *Genetics* 139: 1371-1382.

Hunt, G. J., R. E. Page, Jr., M. K. Fondrk, and C. J. Dullum. 1995. Major quantitative trait loci affecting honey bee foraging behavior. *Genetics* 141: 1537-1545.

Jordan, R. A., and R. W. Brosemer. 1974. Characterization of DNA from three bee species. *J. Insect Physiol.* 20: 2513-2520.

Karl, S. A., and J. C. Avise. 1993. PCR-based assays of Mendelian polymorphisms from anonymous single-copy nuclear DNA: techniques and application for population genetics. *Mol. Biol. Evol.* 10: 342-361.

Kent, R. B. 1988. The introduction and diffusion of the African honeybee in South America. *Yearbook of the Association of Pacific Coast Geographers* 50: 21-43.

Kerr, W., and D. Bueno. 1970. Natural crossing between *Apis mellifera adansonii* and *Apis mellifera ligustica*. *Evolution* 24: 145-148.

Lindauer, M. 1955 Schwarmbienen auf wohnungssuche. *Zeit. vergl. Physiol.* 37: 263-324.

Lobo, J. A. 1995. Morphometric, isozymic, and mitochondrial variability of Africanized honeybees in Costa Rica. *Heredity* 75: 133-141.

Lobo, J. A., M. A. Lama, and M. A. Mestriner. 1989. Population differentiation and racial admixture in the Africanized honeybee (*Apis mellifera* L.). *Evolution* 43: 794-802.

Lobo, J. A., and H. Krieger. 1992. Maximum likelihood estimates of gene frequencies and racial admixture in *Apis mellifera* L. (Africanized honeybees). *Heredity* 68: 441-448.

McMichael, M., and H. G. Hall. 1996. DNA RFLPs at a highly polymorphic locus distinguish European and African subspecies of the honey bee, *Apis mellifera* L. and identify geographic origins of New World honey bees. *Molec. Ecol.* 5: 403-416.

Michener, C. D. 1975. The Brazilian bee problem. *Ann. Rev. Entomol.* 20: 399-416.

Moritz, C., T. E. Dowling, and W. M. Brown. 1987. Evolution of animal mitochondrial DNA: relevance for population biology and systematics. *Ann. Rev. Ecol. Syst.* 18: 269-292.

Nielsen, D., R. E. Page, Jr., and M. W. J. Crosland. 1994. Clinal variation and selection of MDH allozymes in honey bee populations. *Experientia* 50: 867-871.

Oldroyd, B. P., A. J. Smolenski, J. -M. Cornuet, and R. H. Crozier. 1994. Anarchy in the beehive. *Nature* 371: 749.

Oldroyd, B. P., J. -M. Cornuet, D. Rowe, T. E. Rinderer, and R. H. Crozier. 1995. Racial admixture of *Apis mellifera* in Tasmania, Australia: Similarities and differences with natural hybrid zones in Europe. *Heredity* 74: 315-325.

Page, R. E., Jr. 1989. Neotropical African bees. *Nature* 339: 181-182.

Page, R. E., Jr., G. E. Robinson, and M. K. Fondrk. 1989. Genetic specialists, kin recognition and nepotism in honey-bee colonies. *Nature* 338: 576-579.

Quezada-Euan, J. J. G., and D. M. Hinsull. 1995. Evidence of continued European morphometrics and mtDNA in feral colonies of honey bees (*Apis mellifera*) from the Yucatán peninsula, Mexico. *J. Apicultural Res.* 34: 161-166.

Rinderer, T. E. 1986. Africanized bees: the Africanization process and potential range in the United States. *Bull. Entomol. Soc. Amer.* 32: 222-227.

Rinderer, T. E., R. L. Hellmich, R. G. Danka, and A. M. Collins. 1985. Male reproductive parasitism: a factor in the Africanization of European honey-bee populations. *Science* 228: 1119-1121.

Rinderer, T. E., J. A. Stelzer, B. P. Oldroyd, S. M. Buco, and W. L. Rubink. 1991. Hybridization between European and Africanized honey bees in the Neotropical Yucatan peninsula. *Science* 253: 309-311.

Rinderer, T. E., B. P. Oldroyd, and W. S. Sheppard. 1993. Africanized bees in the U.S. *Sci. Amer.* 269: 84-90.

Saiki, R. K., D. H. Gelfand, S. Stoffel, S. J. Scharf, R. Higuchi, and G. T. Horn. 1988. Primer-directed enzymatic amplification of DNA with a thermostable DNA polymerase. *Science* 239: 487-491.

Sheppard, W. S., T. E. Rinderer, J. A. Mazzoli, J. A. Stelzer, and H. Shimanuki. 1991. Gene flow between African- and European-derived honey bee populations in Argentina. *Nature* 349: 782-784.

Smith, D. R. 1991. Mitochondrial DNA and honey bee biogeography. pp. 131-176. *In:* Diversity in the Genus *Apis*. Smith, D.R. [ed.]. Westview Press, Boulder.

Smith, D. R., O. R. Taylor, and W. M. Brown. 1989. Neotropical Africanized honey bees have African mitochondrial DNA. *Nature* 339: 213-215.

Smith, D. R., and W. M. Brown. 1990. Restriction endonuclease cleavage site and length polymorphisms in mitochondrial DNA of *Apis mellifera mellifera* and *A. m. carnica* (Hymenoptera: Apidae). *Ann. Ent. Soc. Amer.* 83: 81-88.

Smith, D. R., M. F. Palopoli, L. Garney, J. -M. Cornuet, M. Solignac, and W. M. Brown. 1991. Geographical overlap of two mitochondrial genomes in Spanish honey bees (*Apis mellifera iberica*). *J. Heredity* 82: 96-100.

Southern, E. M. 1975. Detection of specific sequences among DNA fragments separated by gel electrophoresis. *J. Molec. Biol.* 98: 503-517.

Southwick, E. E., D. W. Roubik, and J. M. Williams. 1990. Comparative energy balance in groups of Africanized and European honey bees: ecological implications. *Comp. Biochem. Physiol.* 97A: 1-7.

Suazo, A., R. McTiernan, and H. G. Hall. 1998. Differences between African and European Honey Bees (*Apis mellifera* L.) using random amplified polymorphic DNA (RAPD). *Journal of Heredity.* 89: 32-36.

Suazo, A., and H. G. Hall. 1999. Detection of nuclear RFLPs in African and European honey bee (*Apis mellifera* L.) DNA amplified with standard and long PCR. In preparation.

Tautz, D. 1989. Hypervariability of simple sequences as a general source for polymorphic DNA markers. *Nucleic Acids Res.* 17: 6463-6471.

Taylor, O. R. 1977. The past and possible future spread of Africanized honeybees in the Americas. *Bee World* 58: 19-30.

Taylor, O. R. 1985a. African Bees: Potential impact in the United States. *Bull. Ent. Soc. Amer.* 31: 14-24.

Taylor, O. R. 1985b. Let's keep our facts straight about the African bees! *Amer. Bee J.* 125: 586-587.

Taylor, O. R. 1988. Ecology and economic impact of African and Africanized honey bees. pp. 29-41. *In:* Africanized Honey Bees and Bee Mites. G. R. Needham, R. E. Page, M. Delfinado-Baker, C. E. Bowman [eds.]. Ellis Horwood Limited, Chichester.

Taylor, O. R., and M. Spivak. 1984. Climatic limits of tropical African honeybees in the Americas. *Bee World* 65: 38-47.

Taylor, O. R., A. Delgado, and F. Brizuela. 1991. Rapid loss of European traits from feral neotropical African honey bee populations. *Amer. Bee J.* 131: 783-784.

Williams, J. G. K., A. R. Kubelik, K. J. Livak, J. A. Rafalski, and S. V. Tingey. 1990. DNA polymorphisms amplified by arbitrary primers are useful as genetic markers. *Nucleic Acids Research* 18: 6531-6535.

Winston, M. L., O. R. Taylor, and G. W. Otis. 1983. Some differences between temperate European and tropical African and South American honeybees. *Bee World* 64: 12-21.

Phylogeny and biogeography of *Apis cerana* subspecies: testing alternative hypotheses

DEBORAH SMITH AND ROBERT H. HAGEN
DEPARTMENT OF ENTOMOLOGY, HAWORTH HALL
UNIVERSITY OF KANSAS, LAWRENCE, KS 66045

e-mail
dsmith@kuhub.cc.ukans.edu
rhagen@kuhub.cc.ukans.edu

Introduction

Phylogenetic analysis has long been used to infer historical biogeography at the level of the species and at higher levels (*e.g.*, Brundin 1965, Funk 1995). Since the mid 1980's, with the development of molecular biological tools it has become possible to construct intra-specific phylogenies that can be used to infer intra-specific biogeography. This extension of the classical approach to historical biogeography has been termed intra-specific phylogeography by Avise *et al.* (1987). The value of intra-specific biogeography is that one can study the processes of adaptive evolution and divergence among populations.

Honey bees are ideal organisms for the study of intra-specific biogeography and phylogeography. First, geographic variation in morphology, behavior, physiology, disease resistance and ecology is well documented for *Apis mellifera* (*e.g.*, Adam 1951, 1954, 1961, 1964, 1977, Ruttner 1988) and the Asian honey bee species (*e.g.*, Maa 1953, Mattu and Verma 1983, 1984a, b, Ruttner 1988, Peng *et al.* 1989, Singh *et al.* 1990, Singh and Verma 1993, Verma *et al.* 1994, Damus 1995, Hadisoesilo *et al.* 1995, Otis 1996).

Second, *Apis* provides a unique opportunity for "replicate studies" in biogeography, particularly in southern Asia. *Apis* comprises three main lineages, the dwarf bees (*A. florea* and *A. andreniformis*), the giant bees (*A. dorsata* in the broad sense), and the cavity-nesting bees (*A. mellifera, A. cerana, A.*

koschevnikovi, and other recently recognized species). The ranges of species in the three lineages overlap broadly in southern Asia. One can compare phylogeographic patterns in each lineage; if the three lineages show identical or nearly identical patterns one could conclude that they responded to external environmental and geological conditions in the same way, and colonized their present ranges

Figure 1. *Apis cerna* collection sites (locations of dots are approximate; see Smith and Hagen (1996) for more precise collection data.

in the same sequence. Alternatively, if each lineage shows a different phylogeographic pattern, it would imply that each lineage colonized its present range in a different sequence of events.

Finally, honey bee populations can provide insight into the relationship between population biology and DNA sequence evolution. For example, *Apis cerana* occurs on the Asian mainland and on islands of varying size in Malaysia, Indonesia, the Philippines and elsewhere, enabling one to investigate the effect of population size on rate of evolution. The islands themselves differ in the length of time since they were last united with the mainland: some, like the island of Luzon, were apparently never connected to the mainland (Heaney 1986) while others, like Borneo and Palawan, were connected to the mainland during the Pleistocene when sea levels were approximately 200 m lower. This makes it possible to investigate length of time a population has been isolated from contact with others as a factor in evolution.

Biogeography of *Apis mellifera*

Among honey bees, modern genetic tools were first applied to the biogeography of *Apis mellifera*. Some of this work grew out of interest in the African-derived bees in the Americas (*e.g.*, Hall 1986, 1988, 1990, 1991, 1992, Smith 1988, 1991a, b, c, Smith and Brown 1988, 1990, Smith *et al.* 1989, 1991, 1997, Hall and Smith 1991, Schiff *et al.* 1994, McMichael and Hall 1996) Other research stemmed from the long-established interest of European honey bee biologists in differentiation among honey

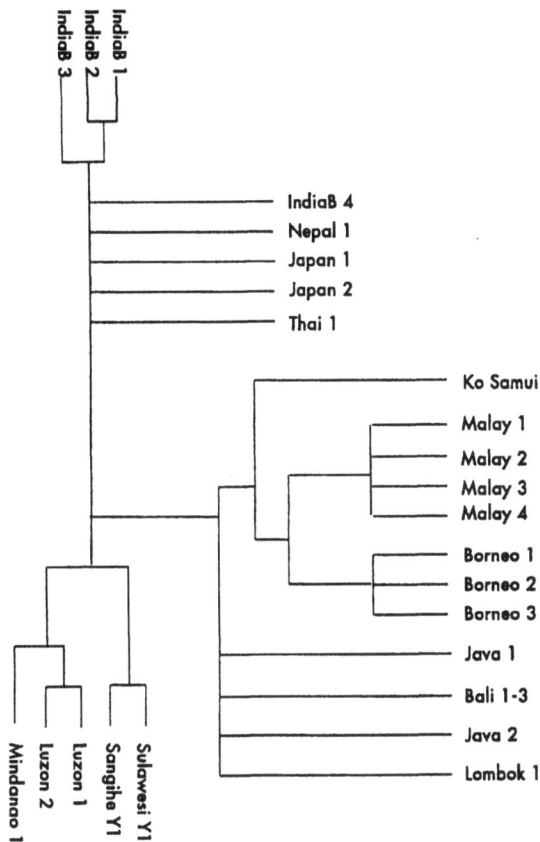

Figure 2. Unrooted network of eastern *Apis cerana* mitochondrial haplotypes. This is a strict consensus of 76 minimal length trees produced by a branch-and-bound search using 30 informative positions in the sequences shown in Figure 3. The haplotype names are the same as in Figure 3.

Table 1. Primers used in amplification and sequencing of non-coding intergenic region of *Apis cerana* mtDNA. Position of primers on *Apis mellifera* mtDNA refers to published sequence of Crozier and Crozier (1993).

Primer sequence (5' to 3')	position of 5' end on *Apis mellifera* mtDNA	Reference
Amplification primers:		
TCTATACCACGACGTTATTC	3090 (Cytochrome Oxidase I)	Hall and Smith 1991
GATCAATATCATTGATGACC	3937 (Cytochrome Oxidase II)	Hall and Smith 1991
Internal sequencing primer:		
GGCAGAATAAGTGCATTG	3363 (leucine tRNA)	Cornuet *et al.* 1991

Table 2.   Comparison of alternative regional groupings of *Apis cerana* populations. The "traditional" grouping follows Ruttner (1988). Alternative 1 combines Japan with *A. c. cerana*. Alternatives 2 and 3 were suggested by the major branches on the cladogram in Figure 2. Alternative 2 maximizes Φ_{ct}, the differentiation among regions.

	Φ_{st}	Φ_{sc}	Φ_{ct}
TRADITIONAL GROUPING:			
A. c. cerana	0.773	0.774	-0.005
Nepal, Hong Kong, Korea			
A. c. indica			
India, Thailand, Malaysia,			
Indonesia, Philippines			
A. c. japonica			
Japan			
ALTERNATIVE 1			
region 1	0.801	0.74	0.221
Nepal, Hong Kong, Korea, Japan			
region 2			
India, Thailand, Malaysia, Indonesia, Philippines			
ALTERNATIVE 2			
region 1	0.796	0.301	0.709
India			
region 2			
Nepal, Thailand, Hong Kong, Korea, Japan			
region 3			
Samui Island, Borneo, peninsular Malaysia			
region 4			
Java, Bali, Lombok, Flores, Timor			
region 5			
Sulawesi			
region 6			
Sangihe			
region 7			
Luzon and Mindanao			
ALTERNATIVE 3			
region 1	0.797	0.597	0.496
India, Nepal, Thailand, Hong Kong, Korea, Japan			
region 2			
Samui Island, Borneo			
peninsular Malaysia			
region 3			
Java, Bali, Lombok, Flores, Timor			
region 4			
Sulawesi, Sangihe, Luzon, Mindanao			

bee races and subspecies (*e.g.*, Garnery *et al.* 1992, 1993, Meixner *et al.* 1993, Moritz *et al.* 1994, Estoup *et al.* 1995, Arias and Sheppard 1996, Sheppard *et al.* 1996).

Mitochondrial DNA (mtDNA) is the most widely used source of information on the phylogeny and biogeography of honey bee races or subspecies. Summarized briefly, studies of mitochondrial DNA restriction site polymorphisms indicate that there are 3 lineages of *A. mellifera* mtDNA. The numerous *A. mellifera* subspecies (24 recognized by Ruttner 1988) can be organized into three major groups, each characterized by possession of one of the three types of mtDNA. The three lineages are: a west European lineage, consisting primarily of *A. m. mellifera*; an east Mediterranean lineage, which includes *A. m. ligustica*, *A. m. carnica*, *A. m. caucasica*, *A. m. anatoliaca*, and probably other subspecies of the eastern Mediterranean and eastern Europe; and an African lineage, which includes *A. m. intermissa*, *A. m. scutellata* and *A. m. capensis* (Smith 1991a, c, Garnery *et al.* 1992, Arias and Sheppard 1996, Smith *et al.* 1997).

The three groups of subspecies—as delineated by their mtDNA types—correspond closely to three groups of subspecies recognized by Ruttner (1988) on the basis of morphometrics. The mitochondrial-based and morphometric-based groupings differ primarily in the treatment of morphologically intermediate populations. For example, morphometric studies show a cline in morphometric characters from *A. m. mellifera* in France, through *A. m. iberica* in Spain and Portugal, *to A. m. intermissa* in north Africa. At first these data were taken to indicate a close relationship between *A. m. mellifera* and the subspecies of north Africa (Cornuet *et al.* 1988, Ruttner 1988, Cornuet and Fresnaye 1989). When mitochondrial DNA data were included in the analysis, they showed that both African and west European mtDNAs occur in Spain (Smith *et al.* 1991; see Franck et al. 1998 for a more detailed discussion of Spanish bees).

Biogeography of *Apis cerana*

As is the case in *A. mellifera*, there is extensive geographic variation among populations of Asian cavity-nesting honey bees (Ruttner 1988). Maa (1953) regarded the Asian cavity-nesting bees a subgenus, *Apis* (*Sigmatapis*), with 11 species and several subspecies. Subsequent authors tended to ignore Maa's classification, and grouped all of the

Figure 3. Observed sequences of the non-coding intergenic region of *A. cerana* mitorial DNA. Each unique sequence is given a number (1-34, first column) and a short name indicating the locality in which it was most common (second column). Only the long eastern sequences are invicated in the AMOVA analysis.

Eastern *A. cerana* haplotypes

```
1)    IndiaB1    AAAATTTAATAAGCTACAATTGCATTAAAT-TATGAATTTAAACTCAAAG--TAAAAAACTTTT-ATT-AAAATTAATAATTTAAATTTATTATTAAAATTT
2)    IndiaB2    AAAATTTAATAAGCTACAATTGCATTAAAT-TATGAATTTAAACTCAAAA--TAAAA--CTTTT-ATT-AAAATTAATAATTTAAATTTATTATTAAAATTT
3)    IndiaB3    AAAATTTAATAAGCTACAATTGCATTAAAT-TATGAATTTAAACTCAAAG--TAAAAA-CTTTT-ATT-AAAATTAATAATTTAAATTTATTATTAAAATTT
4)    IndiaB4    AAAATTTAATAAGCTACAATTGCATTGAAT-TCTAAATTCAAACTCAAAG--TAAAAAACTTTT-ATT-AAAATTAATAATTTAAATTTATTATTAAAATTT
5)    Nepal1     AAAATTTAATAAGCTACAATTGCATTGAAT-TCTGAATTCAAACTCAAAG--TAAAAAACTTTT-ATT-AAAATTAATAATCTAAATTTATTATTAAAATTT
6)    Japan1     AAAATTTAATAAGCTACAATTGCATTGAAT-TCTGAATTCAAACTCAAAG--TAAAAAACTTTT-ATT-AAAATTAATAATTTAAATTTATTATTAAAATTT
7)    Japan2     AAAATTTAATAAGCTACAATTGCATTGAAT-TCTGAATTCAAACTCAAAA--TAAAAAACTTTT-ATT-AAAATTAATAATTTAAATTTATTATTAAAATTT
8)    Korea4     AAAATTTAATAAGCTACAATTACATTGAAT-TCTGAATTCAAACTCAAAG--TAAAAAACTTTT-ATT-AAAATTAATAATTTAAATTTATTATTAAAATTT
9)    Korea7     AAAATTTAATAAGCTACAATTGCATTGAAT-TCTGAATTCAAACTCAAAA--TAAAAAACTTTT-ATT-AAAATTAATAATTTAAATTTATTATTAAAATTT
10)   Korea9     AAAATTTAATAATCTACAATTGCATTGAAT-TCTGAATTCAAACTCAAAG--TAAAAAACTTTT-ATT-AAAATTAATAATTTAAATTTATTATTAAAATTT
11)   Thai1      AAAATTTAATAAGCTACAATTGCATTGAAT-TCTGAATTCAAACTCAAAG--TAAAAA-CTTTT-ATT-AAAATTAATAATTTAAATTTATTATTAAAATTT
12)   KoSamui    AAAATTTAATAAGCTTTAATTGCATTGAAT-TTTAAATTCAAATTCAAAA--TAAAA--CTTTT-ATT-AAAATTAATAATTTAAATTTATTTATTAAAATTT
13)   Malay1     AAAATTTAATAAGCTTTAATTGCATTGAAT-TTTGAATTCAAATTCAAAA--TAAAA-TTTTT-ATT-AAAATTAATAATTTAAATTTATTATTAAAATTT
14)   Malay2     AAAATTTAATAAGCTTTAATTGCATTGAAT-ATTGAATTCAAATTCAAAA--TAAAA-TTTTT-ATT-AAAATTAATAATTTAAATTTATTATTAAAATTT
15)   Malay3     AAAATTTAATAAGCTTTAATTGCATTGAAC-TTTGAATTCAAATTCAAAA--TAAAA-TTTTT-ATT-AAAATTAATAATTTAAATTTATTATTAAAATTT
16)   Malay4     AAAATTTAATAAGCTTTAATTGCATTGAAT-ATTGAATTCAAATTCAAAA--TAAAA-TTTTT-ATT-AAAATTAATAATTTAAATTTATTATTAAAATTT
17)   Borneo1    AAAATTTAATAAGCTTTAATTGCAATGAAT-TTTGAATTCAAATTTAAAA--TAAAA-CTTTT-ATT-AAAATTAATAATTTAAATTTATTATTAAAATTT
18)   Borneo2    AAAATTTAATAAGCTTTAATTGCATTGAAT-TTTGAATTCAAATTTAAAA--TAAAA-CTTTT-ATT-AAAATTAATAATTTAAATTTATTATTAAAATTT
19)   Borneo3    AAAATTTAATAAGCTTTAATTGCATTGAAT-TTTGAATTCAAATTTAAAA--TAAAA-CTTTTTATT-AAAATTAATAATTTAAATTTATTATTAAAATTT
20)   Java1      AAAATTTAATAAGCTATAATTGCATTGAAT-TTTAAATTCAAATTCAAAA--TAAAA-CTTTT-ATT-AAAATTAATAATTTAAATTTATTATTAAAATTT
21)   Bali1      AAAATT-AATAAGCTATAATTGCATTGAAT-TTTAAATTCAAATTCAAAG--TAAAA-CTATT-ATT-AAAATTAATAATTTAAATTTATTATTAAAATTT
22)   Bali2      AAAATTTAATAAGCTATAATTGCATTGAAT-TTTAAATTCAAATTCAAAG--TAAAA-CTATT-ATT-AAAATTAATAATTTAAATTTATTATTAAAATTT
23)   Bali3      AAAATTTATTAAGCTATAATTGCATTGAAT-TTTAAATTCAAATTCAAAG--TAAAA-CTATT-ATT-AAAATTAATAATTTAAATTTATTATTAAAATTT
24)   Java2      AAAATTTAATAAGCTATAATTGCATTGAAT-TTTAAATTCAAATTCAAAA--TAAAA-TTTTT-ATT-AAAATTAATAATTTAAATTTATTATTAAAATTT
25)   Lombok1    AAAATTTAATAAGCTATAATTGCATTGAAT-TTTAAATTCAAATTCAAAA--TAAAA-CTATT-ATT-AAAATTAATAATTTAAATTTATTATTAAAATTT
26)   SulawesiY1 AAAATTTAATAAACTATATTTACATTGAAT-TATAA-TTCAATCCTAAAGTTTATAAA-CTTT--ATT-AAAATTAATAATTTAC--TTATTATTAAAATTT
27)   SangiheY1  AAAATTTAATAAACTATATTTACATTGAAT-TATAA-TTCAATCCTAAAGTTTAAAAA-CTTT--ATT-AAAATTAATAATTTAC--TTATTATTAAAATTT
28)   Luzon1     AAAATTTAATAAAATACAAATACATTGAAT-TATAA-TTCAAAATTAAAGTATAA----CTTT--ATT-AAATTTAATAATTTAA-ATTATTATTAAAATTT
29)   Luzon2     AAAATTTAATAAAATACAAATAAATTGAATCTATAA-TTCAAAATTTAAGTATAA----CTTT--ATT-AAATTTAATAATTTAA-ATTATTATTAAAATTT
30)   Mindanao1  AAAATTTAATAAACTT-AAATATATTGAAT-TTTAAATTCAAACTTAAAATAATA----TTTT--ATTTAAAATTAATAATTTAA--TTATTATTAAAATTT
```

Short eastern *A. cerana* haplotpyes

```
31)   SulawesiS  AAAATT-AATAATTT-AA-TTATTATTAAAATTT
32)   TaiwanS    AAAATT-AATAATTTTAATTTATTATTAAAATTT
```

Western *A. cerana* haplotypes

```
33)   IndiaY1    AAAATTTAATAAATTATAATTATATTTAATTTTTAATATAATATACAATTTTATAATTTATTAAAATTAATAATTTAAAAATTTATTATTAAAATTT
34)   IndiaY2    AAAATTTAATAAATTATAATTATATTTAATTTTTAA-ATAATATACAATTTTATAATTTATTAAAATTAATAATTTAAAAATTTATTATTAAAATTT
```

Asian cavity-nesting bees into one species, *A. cerana*. More recently, several of Maa's species have been recognized as valid and "split off" from *A. cerana* (e.g., *Apis koschevnikovi*; Koeniger *et al.* 1988, Tingek *et al.* 1988, Rinderer *et al.* 1989, Ruttner *et al.* 1989).

Ruttner (1988) recognized 4 subspecies of *A. cerana*: a northern subspecies, *A. c. cerana*, a southern subspecies, *A. c. indica*, and two more localized subspecies: *A. c. japonica* in Japan and *A. c. himalaya* in the Himalayan region. Other authors (e.g., Peng *et al.* 1989) have recognized geographical races within *A. c. cerana* and *A. c. indica*.

In an earlier study (Smith and Hagen 1996) we examined geographical variation in the mtDNA of *A. cerana* collected from several populations (Figure 1). Instead of the mitochondrial DNA restriction site polymorphisms used in our earlier stud-

ies of A. mellifera, we used the DNA sequence of a non-coding intergenic region in the mitochondrial genome (Cornuet *et al.* 1991), and from the aligned sequences constructed an unrooted cladogram of mtDNA haplotypes (Figure 2). Although some haplotypes were widespread and others very localized, this cladogram indicated strong geographic structure in the distribution of mtDNA haplotypes.

In this study, we examine further the phylogeography of these *A. cerana* populations and additional samples from Korea. We employ the analysis of molecular variance (AMOVA) technique as implemented in the computer software, AMOVA (Excoffier *et al.* 1992) to test alternative groupings of *A. cerana* populations: that suggested by Ruttner (1988) and several groupings suggested by the unrooted cladogram of *A. cerana* mtDNA haplotypes (Smith and Hagen 1996).

Methods

Collections. Samples of *A. cerana* workers were obtained from the localities shown in Figures 1 and 4. More detailed information on collection sites is presented in Smith and Hagen (1996). Samples were preserved in 70% ethanol, or frozen in liquid nitrogen and stored in a -80°C freezer. Samples from 19 of these localities (indicated in Figure 4) were used in the analysis of molecular variation discussed below.

Laboratory methods. Genomic DNA was extracted from the thoraces of single bees. A portion of the mitochondrial genome, including the 3' end of the cytochrome oxidase I gene (COI), leucine tRNA gene, non-coding intergenic region, and 5' end of cytochrome oxidase II (COII) was amplified by means of the polymerase chain reaction (PCR; Saiki *et al.* 1985) using the primers shown in Table 1. The non-coding intergenic region was sequenced manually using the fMol cycle sequencing kit (Promega) and the sequencing primer shown in Table 1. More complete protocols for DNA extraction, PCR conditions, primers, and manual sequencing are presented in Smith and Hagen (1996).

Phylogenetic analysis. Sequences of the non-coding region were aligned manually and with the alignment program ClustalW (Higgins and Sharp 1988). The aligned sequences were reduced to a matrix containing only variable sites. The informative characters in this data set were used to construct midpoint-rooted minimal trees of the haplotypes, using a branch and bound search, with all positions in the matrix, including gaps, weighted equally (we used the PAUP computer program, Swofford 1989).

Distance measures. We estimated pairwise sequence divergences among non-coding sequences 1 through 30 (Figure 3) using the Jukes-Cantor 1-parameter model in the computer program MEGA (Kumar *et al.* 1993). We chose the Jukes-Cantor distance since all of our distances are less than 0.30 (30%) and transitions were not more common than transversions (Kumar *et al.* 1993). The MEGA implementation of Jukes-Cantor distance ignores gaps in sequences, either by eliminating a position from analysis if any of the aligned sequences has a gap at that site, or by ignoring only those gaps involved in each particular pairwise comparison. We chose the latter option.

Analysis of molecular variance. Nineteen populations or collection sites (indicted in Figure 4) were

included in our analysis of molecular variance. We used the AMOVA approach to compare several alternative groupings of these populations into regions, as shown in Table 2. We used the cladogram of haplotypes in Figure 2 to suggest groupings of *A. cerana* populations. We also grouped the populations according to the more traditional subspecies designations, *A. c. cerana* (our Nepal, Hong Kong and Korean samples), *A. c. japonica* (our samples from Japan), and *A. c. indica* (all others).

AMOVA is a hierarchical approach that takes into account not only the number and frequency of haplotypes shared among populations, but also the similarity of the haplotype sequences. It generates a set of Φ-statistics, analogous to Wright's F-statistics (Wright 1951). Φ_{st} is analogous to Wright's F_{st}, and measures the correlation among haplotypes drawn at random from the same population, relative to the correlation of two haplotypes drawn from the species as a whole. Φ_{ct} is the correlation among haplotypes drawn at random from a region (group of populations), compared to the correlation among haplotypes drawn at random from the species as a whole. It measures the genetic variation among regions. Φ_{sc} is the correlation among haplotypes drawn at random from within a population, compared to the correlation among haplotypes drawn at random from the same region. It measures the proportion of variation among different populations within a region. Our goal is to group populations in such a way as to maximize Φ_{ct}, the variation among the regional groupings (see, *e.g.*, the study by Stanley *et al.* 1996).

Results

Sequencing revealed 34 different intergenic sequences, shown in Figure 3. Thirty of these sequences (1-30 in Figure 3) could be aligned with one another. Two sequences (33 and 34) differed from sequences 1-30 so much that alignment was difficult: with a preponderance just 2 bases (A and T) and no limit on insertion of gaps, alignment is somewhat arbitrary. In Smith and Hagen (1996) we called the first 30 sequences "the eastern group" and the latter two sequences "the western group". Both eastern and western sequences were found in India. Sequences 31 and 32 are very short: nearly all of the non-coding sequence is absent. These sequences were found in Sulawesi and Taiwan, respectively. Table 3 shows the number and types of sequences, or haplotypes, found in each collection

Figure 4. Geographic distribution of the 34 different non-coding intergenic sequences found in A. cerana. This table shows how many examples of each sequence were found in each population or collection site. Populations numbered 1 through 19 and sequences 1 through 30 were used in the molecular analysis of variance.

Sequences

Population	1	2	3	4	5	6	7	8	9	10	11	12	13	14	15	16	17	18	19	20	21	22	23	24	25	26	27	28	29	30	31	32	33	34
1) India																																		
Bangalore		1	1	2																					3								1	1
Andamans																																		
Sri Lanka																																		
2) Nepal					1																													
3) Thailand																																		
north						2																												
south										3																								
8) Samui Is.											1																							
5) Hong Kong						2																												
Taiwan					1			1	1																					5				
6) Korea						1	7	1																										
7) Japan							14	1																										
9) Malaysia peninsula													5		1	1	1	1	2	1														
10) Borneo																																		
Indonesia																																		
11) Java																7	1				1	1	1											
12) Bali																								4										
13) Lombok																								4										
14) Flores																								5										
15) Timor																								10										
16) Sulawesi																										2	2		4					
19) Sangihe																											6	1						
Philippines																																		
17) Luzon																												1	5	1				
18) Mindanao																																		
Population	1	2	3	4	5	6	7	8	9	10	11	12	13	14	15	16	17	18	19	20	21	22	23	24	25	26	27	28	29	30	31	32	33	34

site or population. Data from other regions of the mitochondrial genome lead us to believe that sequences 31 and 32 belong to the eastern group, and that they represent independent losses of large portions of the non-coding region (Smith and Hagen, 1996 and unpublished data).

Only sequences 1-30, the eastern group, were used in AMOVA. Jukes-Cantor pairwise distances among the 30 eastern sequences ranged from 0 to 0.1765. (Zero values could be obtained since the MEGA implementation of the Jukes-Cantor distance ignores sequence positions with gaps; some sequences differed only by an insertion/deletion event, which is shown in the aligned sequences by a gap in one, a base in the other). The results of the analyses of molecular variance are shown in Table 2. The grouping with the largest value of Φ_{ct} (0.709), and hence, the grouping that maximizes differences among groups while minimizing variation within groups, is (India) (Nepal, north and south Thailand, Hong Kong, Korea, Japan) (Samui Island, peninsular Malaysia, Borneo) (Java, Bali, Lombok, Timor, Flores) (Sulawesi) (Sangihe) (Luzon, Mindanao). The distinctiveness of the Indian population would be accentuated if sequences 33 and 34 were included in the analyses. The traditional grouping of populations into *A. c. cerana*, *A.c. indica* and *A. c. japonica* was unsupported by this mtDNA data (Φ_{ct}= -0.005). However, the Φ_{ct} value is improved by grouping Japan with the other *A. c. cerana* populations (Φ_{ct} = 0.221).

Why do morphometrics and mtDNA give different pictures of *Apis cerana* biogeography? Our groupings differ from the traditional subspecies grouping primarily in (1) the placement of Indian, Thai and Japanese bees and (2) in recognizing more subdivisions. India forms a separate region, due to the variety and distinctiveness of haplotypes found there. The bees of Japan group with other bees of mainland Asia: Nepal, Korea, and Hong Kong. The Thai bees also group with this region, not with bees of Malaysia and Indonesia. The populations traditionally referred to as *A. c. indica* are subdivided into a "Malaysian group" containing Samui island, peninsular Malaysia and Borneo, an "Indonesian group" including bees from Java, Bali, Lombok, Flores and Timor, and a "Philippine group" containing Luzon and Mindanao. The islands of Sulawesi and Sangihe each form separate groups.

The mitochondrial DNA data reveal historical patterns in colonization, isolation and divergence. At various times during the Pleistocene, sea levels were as much as 200 m. lower than at present. Borneo and many islands of the Indonesian archipelago (a grouping referred to as Sundaland) would have been united to the mainland. In particular, the Malaysian, Indonesia, Philippine, Sulawesi and Sangihe regions our study revels are probably units that have been isolated from the mainland and from one another for different lengths of time (Heaney 1986, Smith and Hagen 1996).

Morphometrics most likely tracks recent evolution and adaptation to current environments in addition to more ancient history. Because morphology is subject to direct selection pressure, morphological similarities among populations in similar habitats or at similar latitudes may be due to convergence.

Acknowledgements

We wish to thank all the people who have helped us collect and acquire bees over the years. We are especially grateful to the late Friedrich Ruttner, Nikolaus and Gudrun Koeniger, and Jean-Marie Cornuet for help in collecting European *A. mellifera*, and to Robin Crewe and Chip Taylor for providing bees from South Africa. For our Asian bee collections we are indebted to Masakmi Sasaki and Tadaharu Yoshida for their help in collecting bees from Japan, and especially to Gard Otis and Martin Damus for collections of bees from throughout southeast Asia. This research was supported by a grant from the General Research Fund, University of Kansas to DRS; by National Science Foundation grant BSR-8918932 to Fred Dyer and DRS; and by National Science Foundation grant BSR-8709661 to DRS.

Literature cited

Adam. 1951. In search of the best strains of bees. *Bee World* 32: 49-52.

Adam. 1954. In search of the best strains of bee: second journey. *Bee World*: 35: 193-203, 233-244.

Adam. 1961. The honey bees of the Iberian peninsula. *Bee World* 42: 252-255.

Adam. 1964. In search of the best strains of bee: concluding journeys. *Bee World* 45: 70-83, 104-118.

Adam. 1977. In search of the best strains of bees: supplementary journey to Asia Minor, 1973. *Bee World* 58: 57-66.

Arias, M. C. and W. S. Sheppard. 1996. Molecular phylogenetics of honey bee subspecies (Apis mellifera L.) inferred from mitochondrial DNA sequence. *Molecular Phylogenetics and Evolution* 5(3): 557-566.

Avise, J. C., Arnold, J., Ball, R. M., Bermingham, E., Lamb, T., Neiggel, J. E., Reeb, C. A. and N. C. Saunders. 1987. Intraspecific phylogeography: the mitochondrial DNA bridge between population genetics and systematics. *Annu. Rev. Ecol. Syst.* 18:489-522.

Brundin, L. 1965. On the real nature of transantarctic relationships. *Evolution* 19:496-505.

Cornuet, J. M., Daoudi, A., Mohssine, E. H., and J. Fresnaye. 1988. Etude biométrique de populations d'abeilles marocaines. *Apidologie* 19: 355-366.

Cornuet, J. M. and J. Fresnaye. 1989. Etude biométrique de populations d'abeilles d'Espagne et du Portugal. *Apidologie* 20:93-101.

Cornuet, J.-M., Garnery, L. and M. Solignac. 1991. Putative origin and function of the intergenic region between COI and COII of *Apis mellifera* L. mitochondrial DNA. *Genetics* 128:393-403.

Crozier, R. H. and Y. C. Crozier. 1993. The mitochondrial genome of the honey bee *Apis mellifera*: complete sequence and genome organization. *Genetics* 133:97-117.

Damus, M. S. 1995. A morphometric and genetic analysis of honey bee (*Apis cerana* F.) samples from Malesia: population discrimination and relationships. M.Sc. Thesis, Department of Graduate studies, University of Guelph.

Estoup, A., Garnery, L., Solignac, M., and J. M. Cornuet. 1995. Microsatellite variation in honey bee (*Apis mellifera* L.) populations: hierarchical genetic structure and test of the infinite allele and stepwise mutation models. *Genetics* 140: 679-695.

Excoffier, L., Smouse, P. E., and J. M. Quattro. 1992. Analysis of molecular variance inferred from metric distances among DNA haplotypes: application to human mitochondrial DNA restriction data. *Genetics* 131:479-491.

Franek, P., Garnery, L., Solignac, M., and J.-M. Cornuet. 1998. The origin of west European subspecies of honeybees (*Apis mellifera*): New insights from microsatellite and mitochondrial data. *Evolution* 52:1119-1134.

Funk, V. A. 1995. Cladistic methods. pp. 30-38 in: W. L. Wagner and V. A. Funk, eds., Hawaiian biogeography: evolution on a hot spot archipelago. Smithsonian Institution Press, Washington DC.

Garnery, L., Cornuet, J. M. and M. Solignac. 1992. Evolutionary history of the honey bee *Apis mellifera* inferred from mitochondrial DNA analysis. *Molecular Ecology* 1 (3): 145-154.

Garnery, L., Solignac, M., Celebrano, G. and J. M. Cornuet. 1993. A simple test using restricted PCR-amplified mitochondrial DNA to study the genetic structure of *Apis mellifera* L. *Experientia* 49(11): 1016-1021.

Hadisoesilo, S., Otis, G. W., and M. Meixner. 1995. Two distinct populations of cavity-nesting honey bees (Hymenoptera: Apidae) in South Sulawesi, Indonesia. *Journal of the Kansas Entomol. Soc.* 68:399-407.

Hall, H. G. 1986. DNA differences found between Africanized and European honeybees. *Proc. Natl. Acad. Sci.* 83:4874-4877.

Hall, H. G. 1988. Characterization of the African honey bee genotype by DNA restriction fragments. pp. 287-293 *In* G. R. Needham, R. E. Page, M. Delfinado-Baker, and C. E. Bowman (eds.), Africanized honey bees and bee mites, Ellis Horwood Ltd., Chichester, England.

Hall, H. G. 1990. Parental analysis of introgressive hybridization between African and European honeybees using nuclear DNA RFLPs. *Genetics* 125:611-621.

Hall, H. G. 1991. Genetic characterization of honeybees through DNA analysis. pp. 45-73 *In* M. Spivak, D. J. C. Fletcher and M. D. Breed (eds), The "African" Honey Bee. Westview Press, Boulder, CO.

Hall, H. G. 1992. Further characterization of nuclear DNA RFLP markers that distinguish African and European honeybees. *Arch. Insect Biochem. Physiol.* 19:163-175.

Hall H. G. and D. R. Smith. 1991. Distinguishing African and European honey bee matrilines using amplified mitochondrial DNA. *Proc. Natl. Acad. Sci. USA* 88:4548-4552.

Heaney, L. R. 1986. Biogeography of mammals in SE Asia: estimates of rates of colonization, extinction and speciation. *Biol. J. Linn. Soc.* 28:127-165.

Higgins, D. G. and P. M. Sharp. 1988. CLUSTAL: a package for performing multiple sequence alignment in a microcomputer. *Gene* 73:237-244.

Koeniger, N., G. Koeniger, S. Tingek, M. Mardan, and T. E. Rinderer. 1988. Reproductive isolation by different time of drone flight between *Apis cerana* Fabricius 1793 and *Apis vechti* (Maa 1953). *Apidologie* 19:103-106.

Kumar, S., Tamura, K. and M. Nei. 1993. MEGA: Molecular evolutionary genetics analysis, version 1.01. The Pennsylvania State University, University Park, PA 16802.

Maa, T.-C. 1953. An inquiry into the systematics of the tribus Apidini or honeybees (Hym.). *Treubia* 21:525-640.

Mattu, V. K. and L. R. Verma. 1983. Comparative morphometric studies on the Indian honey bee of the northwest Himalayas 1. tongue and antenna. *J. Apic. Res.* 22:79-85.

Mattu, V. K. and L. R. Verma. 1984a. Comparative morphometric studies on the Indian honey bee of the northwest Himalayas 2. wings. *J. Apic. Res.* 23:3-10

Mattu, V. K. and L. R. Verma. 1984b. Comparative morphometric studies on the Indian honey bee of the northwest Himalayas 3. hind leg, tergites and sternites. *J.*

Apic. Res. 23:117-122.

McMichael, M. and H. G. Hall. 1996. DNA RFLPs at a highly polymorphic locus distinguish European and African subspecies of the honey bee *Apis mellifera* L. and suggest geographical origins of New World honey bees. *Molecular Ecology* 5(3): 403-416

Meixner, M. D., Sheppard, W. S. and J. Poklukar. 1993. Asymmetrical distribution of a mitochondrial DNA polymorphism between 2 introgressing honey bee subspecies. *Apidologie* 24(2): 147-153.

Moritz, R. F.A., Cornuet, J. M., Kryger, P., Garnery, L. and H. R. Hepburn. 1994. Mitochondrial DNA variability in South African honeybees (*Apis mellifera* L.). *Apidologie* 25(2): 169-178.

Otis, G. 1991. A review of the diversity of species within *Apis*. pp. 29-49 *In* Smith, D. R., (ed.). Diversity in the Genus *Apis*. Westview Press, Boulder, CO.

Peng, Y. S., M. E. Nasr and S. J. Locke. 1989. Geographical races of *Apis cerana* Fabricius in China and their distribution. Review of recent Chinese publications and a preliminary statistical analysis. *Apidology* 20:9-20.

Otis, G. 1996. Distributions of recently recognized species of honey bees (Hymenoptera: Apidae; Apis) in Asia. *J. Kans. Entomol. Soc.* 69 (4) suppl., pp. 331-333.

Rinderer, T. E., N. Koeniger, S. Tingek, M. Mardan and G. Koeniger. 1989. A morphological comparison of the cavity dwelling honeybees of Borneo *Apis koschevnikovi* (Buttel-Reepen, 1906) and *A. cerana* (Fabricius, 1793). *Apidologie* 20:405-411.

Ruttner, F. 1988. Biogeography and taxonomy of honeybees. Springer-Verlag, Berlin.

Ruttner, F., D. Kauhausen and N. Koeniger. 1989. Position of the red honey bee, *Apis koschevnikovi* (Buttel-Reepen 1906), within the genus *Apis*. *Apidologie* 20:395-404.

Saiki, R. K., Scharf, S., Faloona, F., Mullis, K. B., Horn, G. T., Erlich, H. A. and N. Arnheim. 1985. Enzymatic amplification of β-globin genomic sequences and restriction site analysis for diagnosis of sickle cell anemia. *Science* 230:1350-1354.

Schiff, N. M., Sheppard, W. S., Loper, G. M. and H. Shimanuki. 1994. Genetic diversity of feral honey bee (Hymenoptera: Apidae) populations in the southern United States. *Annals of the Entomological Society of America* 87(6): 842-848.

Sheppard, W. S., Rinderer, T. E., Meixner, M.D., Yoo, H. R., Stelzer, J. A., Schiff, N. M., Kamel, S. M. and R. Krell. 1996. Hinfl variation in mitochondrial DNA of old world honey bee subspecies. *Journal of Heredity* 87(1): 35-40.

Singh, M. P. and L. R. Verma. 1993. Morphometric comparison of three geographic populations of the northeast Himalayan *Apis cerana*. pp. 67-80 *In* L. J. Connor, T. Rinderer, H. A. Sylvester and S. Wongsiri (eds.), Asian Apiculture. Wicwas Press, Cheshire CT.

Singh, M. P., Verma, L. R., and H. V. Daly. 1990. Morphometric analysis of the Indian honeybee in the northeast Himalayan region. *J. Apic. Res.* 29:3-14.

Smith, D. R. 1988. Mitochondrial DNA polymorphisms in five Old World subspecies of the honey bee and in New World hybrid populations. pp. 303-312 *In* G. R. Needham, R. E. Page, M. Delfinado-Baker, and C. E. Bowman (eds.), Africanized honey bees and bee mites,

Ellis Horwood Ltd., Chichester, England.

Smith D. R. 1991a. African bees in the Americas: Insights from Biogeography and Genetics. *Trends in Ecol. Evol.* 6:17-21.

Smith, D. R. (ed.) 1991b. Diversity in the Genus *Apis.* Westview Press, Boulder CO, 263 pp.

Smith, D. R. 1991c. Mitochondrial DNA and honey bee biogeography. pp. 131-176 *In* Smith, D. R. (ed.), Diversity in the Genus *Apis.* Westview Press, Boulder, CO.

Smith, D. R. and W. M. Brown. 1988. Polymorphisms in mitochondrial DNA of European and Africanized honey bees (*Apis mellifera*). *Experientia* 44:257-260.

Smith, D. R. and W. M. Brown. 1990. Restriction endonuclease cleavage site and length polymorphisms in mitochondrial DNA of *Apis mellifera mellifera* and *A. m. carnica* (Hymenoptera: Apidae). *Ann. Entomol. Soc. of Amer.* 83: 81-88.

Smith, D. R. and R. H. Hagen. 1996. The biogeography of *Apis cerana* as revealed by mitochondrial DNA sequence data. *J. Kansas Ent. Soc.* 69 (4) suppl., pp. 249-310.

Smith, D.R., Palopoli, M. F., Taylor, B. R., Garnery, L., Cornuet, J.-M., Solignac, M. and W. M. Brown. 1991. Geographic overlap of two classes of mitochondrial DNA in Spanish honey bees (*Apis mellifera iberica*). *Journal of Heredity* 82: 96-100.

Smith, D. R., Slaymaker, A., Palmer, M. and O. Kaftanoglu. 1997. Turkish honey bees belong to the east Mediterranean mitochondrial lineage. *Apidologie* 28: 269-274.

Smith, D. R., Taylor, O. R. and W. M. Brown. 1989. Neotropical Africanized honey bees have African mitochondrial DNA. *Nature* 339:213-215.

Stanley, H. F., Casey, S. , Carnahan, J. M., Goodman, S., Harwood, J. and R. K. Wayne. 1996. Worldwide patterns of mitochondrial DNA differentiation in the harbor seal (*Phoca vitulina*). *Mol. Biol. Evol.* 13:368-382.

Swofford, D. L. 1993. PAUP: phylogenetic analysis using parsimony, version 3.1. *Illinois Natural History Survey.*

Tingek, S., M. Mardan, T. E. Rinderer, N. Koeniger and G. Koeniger. 1988. Rediscovery of *Apis vechti* (Maa 1953): the Saban honey bee. *Apidologie* 19:97-102.

Verma, L. R., Mattu, V. K., and H. V. Daly. 1994. Multivariate morphometrics of the Indian honeybee in the northwest Himalayan region. *Apidologie* 25: 203-223.

Wright, S. 1951. The genetical structure of populations. *Annals of Eugenics* 15: 323-354.

Insights into honey bee biology from *Apis nigrocincta* of Indonesia

GARD W. OTIS AND SOESILAWATI HADISOESILO[2]

DEPARTMENT OF ENVIRONMENTAL BIOLOGY
UNIVERSITY OF GUELPH
GUELPH, ONTARIO, CANADA N1G 2W1
Phone: (519) 824-4120
FAX: (519) 837-0442
E-mail: gotis@evbhort.uoguelph.ca

[2]FORESTRY RESEARCH AND DEVELOPMENT AGENCY,
MANGGALA WANABAKTI BLOCK I/XI
JAKARTA 10270, INDONESIA

Summary

We recently rediscovered *Apis nigrocincta* in Sulawesi, Indonesia. Its similarity to *A. cerana* indicates that divergence between these two species occurred more recently than between other species of honey bees. Consequently, this species pair is a model for the study of speciation of honey bees. A behavioral trait, the time of mating, was identified as a very important component of the mate recognition systems of these two species that prevents hybridization between them. In contrast, the male genitalia of these two species differ only slightly, leading to the conclusions that (1) behavioral differences are probably more important than morphological differences in the speciation of honey bees, and (2) lack of distinctive genital form does not preclude the possibility that a population of bees represents a behaviorally distinct species.

In documenting the distribution of these two species, we located one region in which there is a narrow transition zone from *A. nigrocincta* to *A. cerana* that corresponds to an abrupt habitat transition from forest to heavily disturbed agricultural habitats. This suggests that these species have strong habitat preferences, a topic which has not been explored previously with honey bees.

One distinctive feature of *A. nigrocincta* is the capping of its drone cells: it lacks the hard cocoon structure with central pore that is so obvious in *A. cerana*. The apparent loss of this structure in *A. nigrocincta* suggests that these two species differ in some important aspect, but none is yet apparent. A review of the possible functions of the hard cap with pore of *A. cerana* indicates that we do not yet understand the function of this structure.

These discoveries arose from a simple bee survey in 1989, following which we directed research effort toward intriguing observations we made. Our work indicates how chance discoveries can lead to interesting discoveries about honey bees.

Introduction

In 1989, as part of a routine survey of honey bee diversity of the large eastern Indonesian island of Sulawesi (formerly the Celebes; Figure 1; Otis and Hadisoesilo 1990), we made a chance discovery that has proven to be particularly exciting. Maa (1953) had reported that the cavity-nesting honey bees of Sulawesi differed sufficiently to warrant the use of the name *Apis nigrocincta*, first assigned by Frederick Smith in 1861. However, since the early 1900's bee researchers had either indicated that there were no cavity-nesting honey bees on Sulawesi (*e.g.*, Crane 1990) or that the species inhabiting the island was the widespread eastern hive bee, *Apis cerana* (*e.g.*, Gould and Gould 1988; Ruttner 1988). It was therefore of considerable interest that we discovered two forms of cavity-nesting honey bee in South Sulawesi. The smaller and darker form was subsequently determined to be a population of *A. cerana* (Damus 1995; Hadisoesilo *et al.* 1995, Damus and Otis 1997). The larger, yellower form,

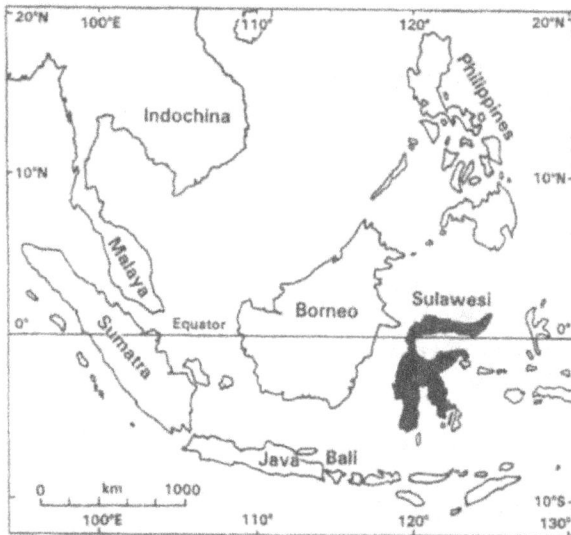

Figure 1. Map of Southeast Asia showing the location of Sulawesi relative to surrounding islands.

although very similar to *A. cerana* and somewhat less similar to *A. koschevnikovi*, was always clearly differentiated in morphometric analyses (Hadisoesilo *et al.* 1995; Hadisoesilo 1997; Damus and Otis 1997) with no intermediates being detected (Figure 2). This suggested that *A. nigrocincta* was a valid species (Otis 1991).

Since 1989, we have had the opportunity to study the honey bee fauna of Sulawesi much more extensively. The results have confirmed the species status of *A. nigrocincta*. Recognition of a new honey bee species is exciting by itself; the genus *Apis*, once believed to contain only four species, is now known to have no less than nine species (Otis 1991; Hadisoesilo and Otis 1996; Tingek *et al.* 1996)! Comparative studies are more interesting now that researchers can study two species of dwarf honey bees, at least two species of giant honey bees, and four species of cavity-nesting honey bees in Asia. In addition, however, the rediscovered species *Apis nigrocincta* presents some situations that offer new insights to the biology of other honey bees.

Apis nigrocincta as a model for speciation in honey bees

In the past, the vast physical and behavioral differences between *A. florea, A. dorsata,* and *A. cerana* prevented speculation on the process of speciation in the genus *Apis*. We do know some relevant pieces of information. For example, the notion that all honey bees evolved from a single common ancestor

(*i.e.*, they form a monophyletic group) has become well supported by morphological, behavioral, and molecular data (Engel and Schultz 1997). There is also agreement that divergence of the dwarf honey bee lineage (*A. florea, A. andreniformis*) from the other honey bees preceded the separation of the giant honey bee clade (*A. dorsata, A. laboriosa*) and the cavity-nesting bee clade (*A. mellifera, A. cerana, A. koschevnikovi, A. nigrocincta, A. nuluensis*) (Engel and Schultz 1997). However, there has been notably little written about how these lineages diverged. The discovery of additional species is providing information that may enable us to fill that void as we uncover pairs of species that are relatively undifferentiated.

To answer the question of "how" different species diverged from their common ancestors, bee biologists must rely on indirect evidence. In the absence of relevant fossils, we can only view the results of evolution, namely living species and their physical and behavioral characteristics, and speculate on the processes that may have occurred. In the case of honey bees, most attention in the past has been directed at what were referred to as "reproductive isolating mechanisms" (Mayr 1963), characteristics which were believed to have evolved to prevent crossing between different populations. In the more modern view of this subject, each species has its own "mate recognition system" that functions to enable the union of queens and drones to produce viable offspring (see Paterson 1985, and references therein). For honey bees, the components of the mate recognition system are identical to premating isolating mechanisms. These are: temporal occurrence of mating flights, location of mating, mating attractants, acceptance or rejection of males by queens, and compatibility of male and female genitalia. In the past, it has never been possible to

Figure 2. *Apis cerana* (left) and *Apis nigrocincta* (right) are extremely similar. *A. nigrocincta* is yellower and usually larger.

determine which of these components is of primary importance at the time that speciation is occurring. In the paragraphs that follow, what is known about these five potential components of the mate recognition system of honey bees is reviewed, following which the new evidence from *Apis nigrocincta* is presented.

Initially, studies focused on the form of the male endophallus. It is common practice for insect systematists to use the form of male genital structures to characterize species, because genital form tends to be species-specific for species with internal fertilization (Shapiro and Porter 1989). The endophalli of the four commonly recognized species of honey bees were found to differ strikingly (Simpson 1960, 1970). As additional populations of bees were studied, their acceptance as species was almost guaranteed if the genitalia of their drones exhibited a characteristic and unique form. Specifically, bee biologists readily accepted *A. koschevnikovi* and *A. andreniformis* as species on the basis of their distinctive male genitalia (Tingek *et al.* 1988; Wongsiri *et al.* 1990). In contrast, the endophallus of *A. laboriosa* differs only slightly, if at all, from that of *A. dorsata* (McEvoy and Underwood 1988; Koeniger *et al.* 1990). Consequently, *A. laboriosa* has only gradually gained acceptance as a valid species, and in the absence of distinct genital characters some scientists still contend that it should continue to be viewed as a subspecies of *A. dorsata*.

Many bee researchers once believed that the male endophallus must fit into the queen's vagina like a "key" in a "lock", and lack of proper fit would prevent hybridization (Eberhard 1985 1990; Shapiro and Porter 1989). Although incompatibility of genital structures may be potentially important between the three main groups of honey bees (dwarf, giant, and cavity-nesting clades), it is likely of little importance for more closely related species. For example, as early as 1983, Ruttner and Maul demonstrated that queens of *A. cerana* can physically mate with drones of *A. mellifera*. Between the Asian species of cavity nesting honey bees whose genital form is even more similar than between *A. mellifera* and *A. cerana*, it is unlikely that the slight differences in form can physically prevent copulation.

Temporal segregation of mating, another component of the mate recognition system of honey bees, was first explored by Koeniger and Wijayagunasekera (1976). They demonstrated that each of the three species in Sri Lanka had a distinct mating flight period. Similar studies have been conducted subsequently in other localities with different complexes of sympatric species: Borneo (Koeniger *et al.* 1988, 1996), Thailand (Rinderer *et al.* 1993), Japan (Yoshida *et al.* 1994), and Peninsular Malaysia (G. Otis and A. Zainal, pers. obs.). In general, mating flight periods of sympatric species overlap only slightly, if at all. The sole known exception is *A. florea* and *A. cerana* in Thailand. These results suggest that timing of mating flights is an important component of the mate recognition systems of different honey bee species.

More recently, it has been discovered that the location of mating also varies between species. It is generally believed that drones and queens mate at drone congregation areas (DCAs) (Zmarlicki and Morse 1963; Koeniger 1991). Where they have been documented, DCAs of Asian honey bee species are located in different places. Drones of *A. cerana* assemble in clearings below the canopy of the trees (Punchihewa *et al.* 1990) or in open areas surrounded by low trees (Yoshida 1994). Drones of *A. dorsata* congregate below the crowns of tall emergent trees (Koeniger *et al.* 1994). Although no DCAs have been located yet, the drones of *A. koschevnikovi* have been observed flying near ground level below the forest canopy (N. Koeniger, pers. comm.). Unfortunately, no study has yet documented the locations of DCAs for all the sympatric species in one location. As a result, the relative importance of the location of mating cannot yet be fully evaluated as a component of the mate recognition system.

Queen mandibular pheromone serves as a mating attractant in honey bees (reviewed by Winston and Slessor 1992). QMP blends differ qualitatively and quantitatively among queens of different species (Plettner *et al.* 1997). However, there is no evidence that these differences function to enhance the intraspecific encounter rate of queens and drones. In fact, drones of all species that have been tested (*A. mellifera*, *A. cerana*, *A. dorsata*) respond to 9-ODA alone (Butler and Fairey 1964; Punchihewa *et al.* 1990; Koeniger *et al.* 1994; Yoshida 1994). (We note that *A. florea*, whose drones have not been tested, has a QMP blend that differs markedly from that of queens of all other species, and its mandibular glands contain very little 9-ODA; Plettner *et al.* 1997). The relative importance of queen pheromones as a component of the *Apis* mate recognition system remains unknown.

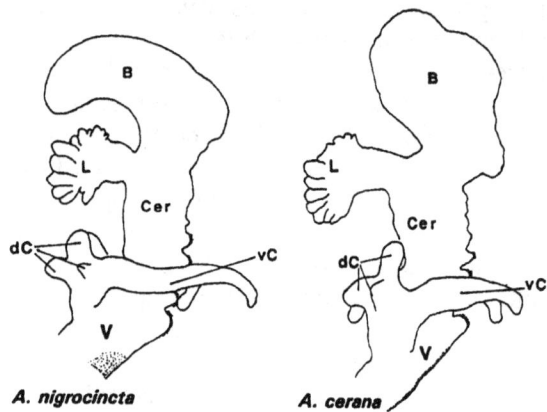

Figure 3. Drawings of everted endophalli of A. nigrocincta and A. cerana made from photographs. B = bulbus; L = fimbriate lobe; vC = ventral cornu; dC = dorsal cornua; Cer = cervix; V = vestibulum.

There is presently no evidence for any species of honey bee to indicate that queens select or reject individual drones as mates. Rather, males compete with other males as they chase queens, and those that successfully reach the queen and grasp her probably mate successfully. The queen must open her sting chamber to allow mating to occur, however, so potentially she could reject some males at that time.

Now, having reviewed what is known about the mate recognition systems of honey bees in general, we summarize what we have learned about *Apis nigrocincta*. This species is so similar to *A. cerana* that no diagnostic characters of adult or larval workers have been identified that enable positive identification (M. Engel, pers. comm.), other than color of the hind femur and clypeus (which is sufficient for Sulawesi specimens, but not across the full range of *A. cerana*). Even in the first report, which drew attention to this possible species, it was noted that the male genitalia of the two morphotypes of Sulawesi bees could not be differentiated (Otis 1991). This result has been confirmed by detailed inspection of both everted and uneverted endophalli (Hadisoesilo 1997). Only minor differences in genital form (*e.g.*, length of the dorsal cornua, length and width of ventral cornua, number of papillae of the fimbriate lobe) have been found (Figure 3).

In contrast to the similarity of the physical characteristics of these two species, timing of mating flights differs greatly (Hadisoesilo and Otis 1996).

Otis & Hadisoesilo: Apis nigrocincta of Indonesia

In each of three locations in Sulawesi, the drones of *A. cerana* had nearly completed their mating flights before the first orientation flights of *A. nigrocincta* occurred (Figure 4). This behavioral trait ensures intraspecific encounters of drones and queens. Conversely, it severely limits the potential for hybridization. In fact, Hadisoesilo (1997) conducted morphometric analyses of samples of bees (10 bees each) from 126 colonies, 51 of which were from the two identified zones of sympatry. Every bee examined was unambiguously identified, and the identification matched the preliminary identification based on leg coloration. Bees with intermediate morphology, which would indicate hybridization, were notably absent from these samples. Unfortunately, we have no data yet on the location of DCAs or the queen mandibular pheromone of *A. nigrocincta* to compare with similar data for *A. cerana*.

The conclusion from these studies is that *A. nigrocincta* and *A. cerana*, despite their similarity, are functioning as two distinct species. Whereas distinctive genitalia can be used to confirm species

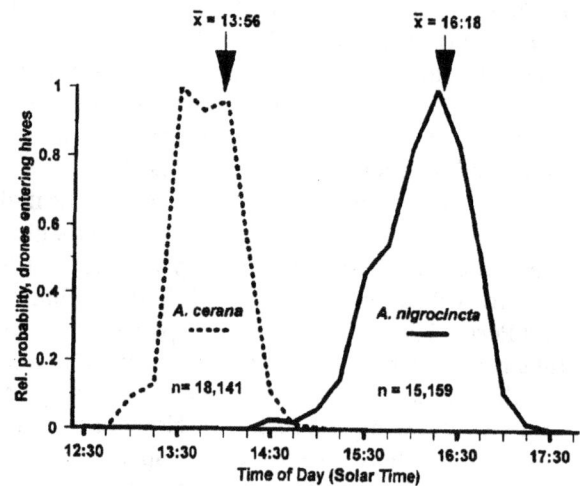

Figure 4. Temporal segregation of mating flights by drones of A. cerana (2 colonies) and A. nigrocincta (3 colonies). Data from a zone of sympatry (Bontobulaeng) obtained over 15 days in September, 1995, were pooled. Sample sizes indicate the number of drones observed entering the colonies; individual bees contributed multiple times to the data set. The data are presented as the proportion of drone flights ending during each 15 min period relative to the maximal number recorded for that species during any 15 min period. Arrows indicate mean drone entrance times. Data are redrawn from Hadisoesilo and Otis, 1996.

status in honey bees (*e.g.*, *A. koschevnikovi*, *A. andreniformis*), the converse is not true: the absence of genital differences cannot be used to demonstrate conspecific status (*e.g.*, *A. laboriosa*, *A. nigrocincta*). It seems that too much importance has been placed on the physical form of drone genitalia as a means of confirming additional species of honey bees (*cf.*, Ruttner 1988). Fruitful lines of investigation would involve documenting the locations of DCAs of additional species, particularly of all sympatric species in single localities (this behavioral trait could vary intraspecifically), and the relative attractiveness of drones to intraspecific and interspecific QMP blends (extracts from queens or synthetic mixtures).

If we wish to infer which biological attributes are most important to the process of speciation, then we should compare the traits of those populations which are most similar and have probably diverged most recently (*i.e.*, are closest to the incipient speciation conditions). In the case of honey bees, the most similar pairs of species are *A. nigrocincta* and *A. cerana* of Sulawesi and *A. nuluensis* and *A. cerana* of Borneo (Tingek *et al.* 1996). Of those aspects of the mate recognition systems of these species pairs that have been studied, behavioral differences (the temporal pattern of mating flights) are much more important than genital differences. Moreover, the timing of mating flights is not constant across all populations of a single species. For example, in Sri Lanka, drones of *A. cerana* take mating flights between 1545 h and 1715 h (Koeniger and Wijayagunasekera 1976). In contrast, in Sulawesi, where *A. nigrocincta* drones are flying during this time interval (*e.g.*, 1500 h to 1715 h), the flight period of *A. cerana* occurs earlier, from 1300 h to 1500 h (Hadisoesilo and Otis 1996). The logical inference from this situation is that the factors (genes?) that control the timing of mating flights can be readily shaped by natural selection to eliminate overlap in drone flight distributions. This topic will be explored in a future publication (Otis *et al.*, in prep.).

The recency of speciation of *A. nigrocincta* and *A. cerana* (*i.e.*, similarity in morphology) makes this species pair a model for studying the process of speciation in honey bees. The same argument can be made for *A. nuluensis* of Borneo. Continued study of these two systems should yield further insights into the evolution of honey bees and their mating systems.

Figure 5. Rain forest habitats of Lore Lindu National Park are separated from heavily disturbed habitats by Jalan Jepang ("Japan Road") near Rahmat, Central Sulawesi.

Habitat preferences of honey hees

Most beekeepers and scientists who work with honey bees have obtained their experience with *Apis mellifera* or *A. cerana*. One reason for this is their manageability for experimentation, honey production, and pollination of crops. Another is the extreme adaptability of both species to different climatic regions and habitat types. Both can be found from sea level to more than 3000 m (10,000 ft) in mountainous regions; from extremely dry deserts to rain forests receiving more than 4000 mm (160 in) of rain per year; from disturbed habitats to pristine natural settings. One of us (GWO) has personally located feral colonies of European races of *A. mellifera* in sugar maple forests of southern Ontario, cypress swamps of Florida, the prairies of Kansas, desert scrub of southern Texas, sculpted sandstone deserts of Utah, and several North American cities.

It is this remarkable diversity of habitats suitable to honey bees that has dulled our senses to possible preferences they may have. Everyone knows that colonies of bees placed in some locations predictably store more honey than those in other apiaries. Moreover, bees seem to winter better in some areas than in others. However, no one has done comparative studies to compare the reproductive rates and survival of bees in different ecological settings, probably because of the amount of work involved. Consequently, we typically have only vague ideas of the numbers of bee colonies inhabiting different regions and the factors contributing to those numbers (*e.g.*, nectar and pollen resources, predators, availability of nest sites, etc.). We also remain completely ignorant of habitat preferences of swarms. For example, despite a large number of studies that have examined the factors that influence acceptance of nest sites (*e.g.*, height above ground, cavity volume, entrance size and orientation; reviewed by Seeley 1985; Winston 1987), none has examined characteristics of the habitat as criteria for the selection of nest sites by swarms. We believe that this reflects a general perception that honey bees do not have strong habitat preferences.

Our work in Sulawesi has caused us to accept that habitat preference can be a very important aspect of the biology of honey bees. We only gradually became aware that the distributions of *A. cerana* and *A. nigrocincta* can be constrained by habitat. Our initial research was conducted in the southern part of South Sulawesi. On this peninsula, we located *A. cerana* colonies in cities (*e.g.*, Ujung Pandang) and disturbed coastal areas. In contrast, only *A. nigrocincta* was found in forested areas and at higher elevations. In the southeastern part of this peninsula (Bontobulaeng, 400-450 m), we found a region in which both species live. There were no obvious differences in the nest sites of the two species (although this needs to be confirmed) in this zone of sympatry, but there was a transition from *A. cerana* to *A. nigrocincta* from lower (southeastern) to higher (northwestern) elevations within this township (Hadisoesilo 1997).

Figure 6. Research assistant, Sugiyani, prepares to sample an *A. cerana* colony established in a storage box inside a house in Kamarora. Numerous colonies like this were reported to us by local residents.

Otis & Hadisoesilo: Apis nigrocincta of Indonesia

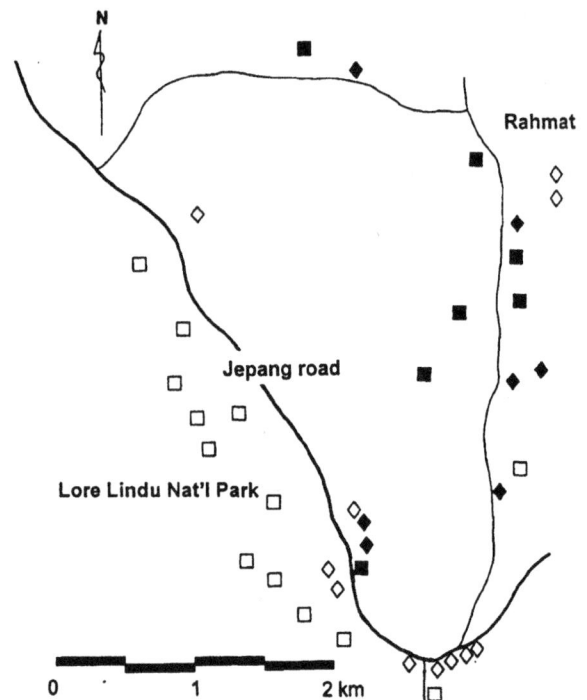

Figure 7. Honey bee colonies located in the vicinity of Rahmat, Central Sulawesi. Open symbols denote colonies of *A. nigrocincta*; solid symbols indicate colonies of *A. cerana*. Colonies located in 1995 and 1996 are indicated by squares and diamonds, respectively. Note the restriction of *A. cerana* to village and agricultural habitats north of Jalan Jepang, and presence of only *A. nigrocincta* in forested habitats of the national park.

It was only when we began to work in the vicinity of Lore Lindu National Park in Central Sulawesi that we became aware of the striking segregation of these two species by habitat (Hadisoesilo 1997). Along the northern boundary of the park in the Palolo valley lie several villages bordered by rain forest. The boundary between rain forest (with understorey partially disturbed by cultivation of coffee, cacao, and several other crops) and disturbed habitats (village, corn fields, plantings of coffee and cacao under small shade trees) is a road (Jalan Jepang; Figure 5). We used two techniques to determine the distribution of bees in this region. We interviewed villagers who led us to colonies in hollow trees, primitive hives, and other cavities (Figure 6). In addition, we sprayed diluted honey on vegetation and recorded the species that foraged on these baits.

Our results were unanticipated. In the vicinity of Kamarora, we located 27 nests in the village itself; 24 of these were of *A. cerana*. In contrast, we recruited thousands of *A. nigrocincta*, but only one *A. cerana*, to honey baits sprayed at 15 forest sites only 0.3-0.8 km from Jalan Jepang. Although it was difficult to locate colonies in the forest, the two we found were of *A. nigrocincta*. Six additional colonies of *A. nigrocincta* were found along a partially forested stream or near the park boundary. In the nearby village of Rahmat, the use of hollowed out sections of coconut trunk as bait hives and the assistance of an informant who was knowledgeable about bees allowed us to locate many more colonies in the forested area. All the colonies in the forested area were of *A. nigrocincta* (n = 19). Across the road in disturbed habitats, most of the colonies located (14/19) were of *A. cerana* (Figure 7).

These data demonstrate a strong association of *A. cerana* with disturbed areas in Central Sulawesi. In contrast, *A. nigrocincta* is abundant in the forested areas and only sporadically found in disturbed sites. We know of no other situation where the micro-distributions of two species of honey bees are so markedly distinct. This raises numerous questions about how bees in general perceive their environment. With respect to this Sulawesi situation, it would be of interest to know the extent to which foragers and scouts from swarms selectively search in specific habitats for food or nest sites, respectively. Does the restriction of *A. cerana* to disturbed habitats indicate a real preference for those habitats, a rejection of cavities in forests, or other factors including interactions with *A. nigrocincta* that result in the observed pattern?

Our discovery of distinct habitat associations of *A. nigrocincta* and *A. cerana* in Central Sulawesi has led us to review other experiences in Southeast Asia. For example, *A. andreniformis* is abundant in peatland areas of western Malaysia that have been converted to coconut plantations with smaller plantings of pineapples, coffee, bananas, and other fruits. In contrast, in the area surrounding the Universiti Pertanian Malaysia about 20 km south of Kuala Lumpur, one of us (GWO) was unable to locate any colonies of this species over a four month period although this species is known to occur on the university lands. Similarly, specimens of *A. koschevnikovi* were collected from the vicinity of Kuala Lumpur earlier in this century, but the only recent collections are from the Pasoh

Forest Reserve several hours to the south. Salmah *et al.* (1990) have documented the habitat associations of social bees in Central Sumatra, and have shown that numbers of honey bees are usually greater in primary forests than in disturbed areas. It is well recognized that human activities affect densities of honey bees, even resulting in local extirpation in some instances (Verma 1993). To what extent are these patterns a function of actual habitat preferences of bees rather than of factors (*e.g.*, habitat alterations, loss of resources, usage of pesticides) that influence birth, growth, and death rates of colonies? With heightened concern about the conservation of biodiversity, this will become a more important avenue of investigation.

The pore in the pupal cappings of drone cells

One of most obvious features of a colony of *A. cerana* is the pore in the cappings of drone cells. After the drone cell containing a large larva is sealed with beeswax, the pupa spins a cocoon which forms a hard conical cap. Subsequently, the pupa apparently secretes a substance that dissolves a hole into the pupal cap, and the workers chew away the beeswax capping leaving the pupal cap with pore exposed (Figure 8) (Sakagami 1960; Hänel and Ruttner 1985). When the drone is mature, it cuts around the edge of the cell capping and emerges. If for some reason the drone dies, it seems that the worker bees rarely remove the cap and the pupal

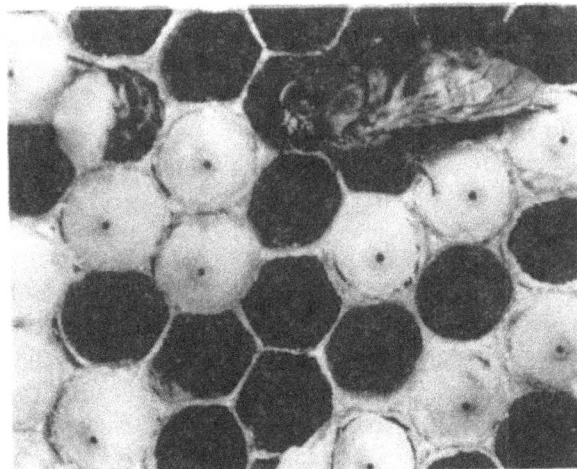

Figure 8. Drone cells of *A. cerana* exhibit the outer portion of the cocoon which consists of a hardened, conical cap with a pore in the center. The outer wax cap has been removed from these cells by workers.

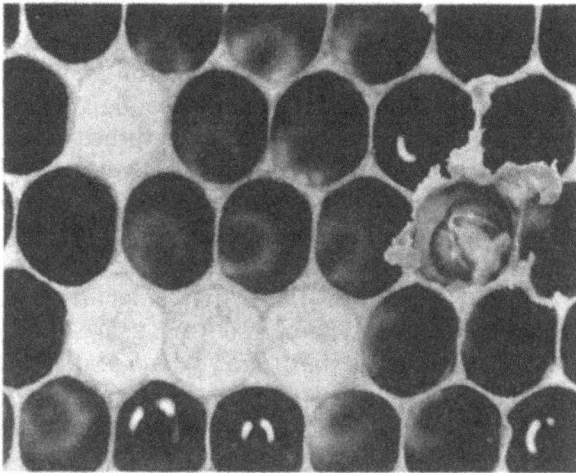

Figure 9. Drone cells of *A. nigrocincta* are sealed with a thin beeswax capping and fragile cocoon. These have been removed from one cell to expose the pupa and demonstrate the absence of the conical cocoon cap with pore which is evident in *A. cerana* (Figure 8).

remains become entombed in the cell. The hardened cell caps with pores are also present in *A. koschevnikovi*.

We were surprised to discover that *A. nigrocincta*, a species very closely related to *A. cerana*, lacks this hard capping of the drone cells. Drone pupae in this species are sealed in by a thin capping of wax and the cocoon which is thin and easily removed (Figure 9). Their cappings resemble those of *A.mellifera*.

This absence of the hard pupal capping with a central pore in *A. nigrocincta* is presently unexplained (Hadisoesilo and Otis 1998). Studies of the biogeography of the bees in this region suggest that *A. nigrocincta* evolved from the Philippine population of *A. cerana* (Damus 1995; Smith and Hagen 1996). As far as is known, all populations of *A. cerana* exhibit the hardened cappings with pores on their drone cells. That this trait apparently has been lost from the Sulawesi population (*A. nigrocincta*) suggests that *A. cerana* and *A. nigrocincta* must differ in some fundamental way. However, our attempts to determine what this may be have been unsuccessful. For example, at lower elevations the two species exist under similar temperature regimes, so climate does not seem to be an important factor. *Varroa jacobsoni* and *V. underwoodi* have been recovered from drone brood of both species, so simple presence or absence of parasitic mites does not provide an explanation. (

Verification of successful reproduction of parasitic mites on *A. nigrocincta* is needed, however.) It is always possible that parasitic mites were absent from Sulawesi until recently, but there is nothing to suggest that this is the case (*e.g.*, *A. nigrocincta* is little affected by its Varroa parasites, which differs from situations in which Varroa has only recently been introduced to a honey bee population). Rath (1992) suggested that drone brood of *A. cerana* produces insufficient amounts of brood pheromones, and that the hardened cell cap evolved to protect the brood from removal by worker bees. This explanation seems unlikely because it requires an increase in frequency of a trait (reduced production of brood pheromone) that actually reduces the fitness of the drones that possess it. Perhaps further study of *A. nigrocincta* and *A. cerana* will clarify this situation as well as lead to a new insights into the biology of drone bees.

Conclusions

Apis nigrocincta is a relatively unimportant species from an economic standpoint. It is found only on Sulawesi and neighboring islands (Otis 1996); recent observations determined that it does not live on Mindano. It is insignificant in terms of honey production, because most honey in Sulawesi is obtained from unmanaged colonies of *A. dorsata binghami*. No one yet manages colonies of *A. nigrocincta* to enhance pollination of crops, such as coffee, fruit trees and coconuts that it visits. Its abundance in forests suggests that it plays a key role in forest ecosystems, but in as much as no one has studied pollination systems in Sulawesi forests to any substantial extent, this remains unknown at present. For most people, *A. nigrocincta* will remain a minor species that is difficult to differentiate from *A. cerana*.

However, as we have described in this chapter, it is this very similarity to *A. cerana* that makes *A. nigrocincta* particularly interesting. Never before have we known of a situation where two species of *Apis* that are nearly ecological equivalents (similar size and morphology) have such strikingly disjunct distributions. Nor have we had the opportunity to get meaningful "snapshots" of the speciation process in honey bees. *A. nigrocincta* will aid in the discovery of additional aspects of bee biology as it becomes better studied.

Before closing, one additional situation deserves

comment. We have already detailed the habitat associations of the bees in the villages of Rahmat and Kamarora on the north edge of Lore Lindu National Park. In the course of performing morphometric analyses on bee samples collected in the forests and disturbed habitats in these villages and surroundings, we discovered that we were working with three, not two, distinct populations of bees! In a plot of discriminant functions 1 and 2 obtained from discriminant analyses, *A. nigrocincta* bees (*n* = 10 samples) formed a tight cluster well separated from *A. cerana* (Figure 10). However, the samples of *A. cerana* from Kamarora (*n* = 6 samples) plotted separately from those collected in Rahmat and Lolu, a village close to Palu (*n* = 8) (Hadisoesilo 1997). Amazingly, Kamarora and Rahmat are only 8 km apart and are very similar ecologically (*e.g.*, climate, habitats, and crops). The biogeographic affinities of these bees are not yet clear, but some biological implications from these results are. Either these two populations of *A. cerana* have only recently arrived in these villages and have not yet had a chance to disperse and interbreed, or there may be three, very similar populations of cavity-nesting honey bees that are all behaving as distinct species in this region! This exciting possibility opens up potentially better opportunities for the study of incipient speciation events in honey bees.

We initiated our studies with a collection of honey bees from South Sulawesi. Our observations from that survey led to the speculation that *Apis nigrocincta* represented an unrecognized species. Subsequent research has upheld that view, but has also led to other, unanticipated avenues of study that provide insights into the basic biology of honey bees. Directed research, in which we stayed within the confines of the known taxonomy and evolutionary views of honey bees, would never have allowed us to make these discoveries. Collectively, our studies summarized here underscore how much we yet have to learn about the genus *Apis* in Southeast Asia, and make a strong argument in favor of basic studies of honey bee biology.

Acknowledgments

We are indebted to Sugiyani, Fitrah Halim, and A. Rahim for their assistance in the field in Sulawesi. The Ujung Pandang office of the Indonesian Ministry of Forestry provided logistical support for the work done in South Sulawesi. Marty

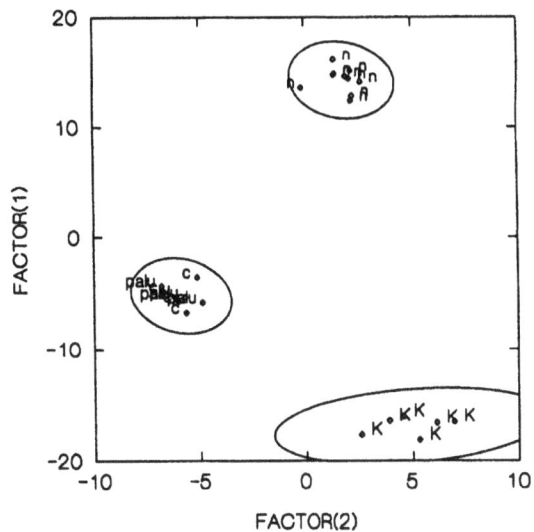

Figure 10. Plot of discriminant functions 1 and 2 for samples of bees collected in Central Sulawesi. Confidence ellipses (95%) are also shown. Collection localities: n = *A. nigrocincta* from Rahmat and Kamarora, Palolo Valley; K = *A. cerana* from Kamarora; c = *A. cerana* from Rahmat; palu = *A. cerana* samples from Lolu, a coconut-growing village near Palu. *A. cerana* from Kamarora and Rahmat, villages only 8 km apart, were classified into two distinct clusters, suggesting they may be from non-interbreeding populations.

Fujita, Wayne Klockner, Duncan Neville, and the staff of The Nature Conservancy in Indonesia facilitated our work in Central Sulawesi. Financial assistance was provided by the Forestry Research and Development Agency of Indonesia, and the Biodiversity Support Program, a consortium of the World Wildlife Fund, The Nature Conservancy, and the World Resources Institute, with funding by the United States Agency for International Development. The opinions expressed herein are those of the authors and do not necessarily reflect the views of the US Agency for International Development or the Nature Conservancy.

References Cited

Butler, C. G. and E. M. Fairey. 1964. Pheromones of the honeybee: biological studies of the mandibular gland secretion of the queen. *J. Apic. Res.* 3: 65-76.

Crane, E. 1990. Bees and Beekeeping. Cornell Univ. Press, Ithaca, NY.

Damus, M. S. 1995. A morphometric and genetic analysis of honey bee (*Apis cerana* F.) samples from Malesia: population discrimination and relationships. M.Sc. thesis, University of Guelph, Ontario, Canada, 119 pp.

Damus, M. S. and G. W. Otis. 1997. A morphometric analysis of *Apis cerana* F and *Apis nigrocincta* Smith populations from Southeast Asia. *Apidologie* 28: 309-323.

Eberhard, W. G. 1985. Sexual Selection and Animal Genitalia. Harvard University Press, Cambridge, MA.

Eberhard, W. G. 1990. Animal genitalia and female choice. *American Scientist* 78: 134-140.

Engel, M. S. and T. R. Schultz. 1997. Phylogeny and behavior in honey bees (Hymenoptera: Apidae). *Ann. Entomol. Soc. Am.* 90: 43-53.

Gould, J. L. and C. G. Gould. 1988. The Honey Bee. W. H. Freeman and Co., New York.

Hadisoesilo, S. 1997. A comparative study of two species of cavity-nesting honey bees of Sulawesi, Indonesia. Ph.D. thesis, University of Guelph, Ontario, Canada, 199 pp.

Hadisoesilo, S. and G. W. Otis. 1996. Drone flight times confirm the species status of *Apis nigrocincta* Smith, 1861 to be a species distinct from *Apis cerana* F, in Sulawesi, Indonesia. *Apidologie* 27: 361-369.

Hadisoesilo, S. and G. W. Otis. 1998. Differences in drone cappings of *Apis cerana* F., 1793 and *Apis nigrocincta* Smith, 1861. *J. Apic. Res.* 37: 11-15.

Hadisoesilo, S., G. W. Otis, and M. Meixner. 1995. Two distinct populations of cavity-nesting honey bees (Hymenoptera: Apidae) in South Sulawesi, Indonesia. *J. Kansas Entomol. Soc.* 68: 399-407.

Hänel, H. and F. Ruttner. 1985. The origin of the pore in the drone cell capping of *Apis cerana* Fabr. *Apidologie* 16: 157-164.

Koeniger, G. 1991. Diversity in Apis mating systems. In: Smith, D. R. (Ed.), Diversity in the Genus *Apis*, pp.199-212. Westview Press, Boulder, CO.

Koeniger, G., M. Mardan, and F. Ruttner. 1990. Male reproductive organs of *Apis dorsata. Apidologie* 21: 161-164.

Koeniger, N., G. Koeniger, M. Gries, S. Tingek, and A. Kelitu. 1996. Reproductive isolation of *Apis nuluensis* (Tingek, Koeniger and Koeniger 1996) by species specific mating time. *Apidologie* 27: 353-360.

Koeniger, N., G. Koeniger, S. Tingek, A. Kalitu, and M. Mardan. 1994. Drones of *Apis dorsata* (Fabricius 1793) congregate under the canopy of tall emergent trees in Borneo. *Apidologie* 25: 249-264.

Koeniger, N., G. Koeniger, S. Tingek, M. Mardan, and T. E. Rinderer. 1988. Reproductive isolation by different time of drone flight between *Apis cerana* Fabricius, 1793 and *Apis vechti* (Maa, 1953). *Apidologie* 19: 103-106.

Koeniger, N. and N. H. P. Wijayagunasekera. 1976. Time of drone flight in three Asiatic honey bee species (*Apis cerana, Apis florea, Apis dorsata). J. Apic. Res.* 15: 67-71.

Maa, T. C. 1953. An inquiry into the systematics of the tribus Apidini or honeybees (Hymenoptera). *Treubia* 21: 525-640.

Mayr, E. 1963. Animal Species and Evolution. Harvard University Press, Cambridge, MA.

McEvoy, M. V. and B. A. Underwood. 1988. The drone and species status of the Himalayan honey bee, *Apis laboriosa* (Hymenoptera: Apidae). *J. Kansas Entomol. Soc.* 61: 246-249.

Otis, G. W. 1991. A review of the diversity of species within Apis. *In:* Smith, D. R. (Ed.), Diversity in the Genus *Apis*, pp. 29-49. Westview Press, Boulder, CO.

Otis, G. W. 1996. Distributions of recently recognized species of honey bees (Apis spp.) in Asia. *J. Kansas Entomol. Soc. Suppl.* 69: 311-333.

Otis, G. W. and S. Hadisoesilo. 1990. Honey bee survey of South Sulawesi. *J. Penelitian Kehutanan* 4: 1-3.

Otis, G. W., S. Hadisoesilo, N. Koeniger, T. E. Rinderer, and M. Mardan. (in prep.) Character divergence in mating times of sympatric species of honey bees (Hymenoptera: Apidae).

Paterson, H. E. H. 1985. The recognition concept of species. In: Vrba, E. S. (Ed.), Species and Speciation, pp. 21-29. Transvaal Museum Monograph No.4. Transvaal Museum, Pretoria.

Plettner, E., G. W. Otis, P. D. C. Wimalaratne, M. L. Winston, K. N. Slessor, T. Pankiw, and R. W. K. Punchihewa. 1997. Species- and caste-determined mandibular gland signals in honey bees (*Apis*). *J. Chem. Ecol.* 23: 363-377.

Punchihewa, R. W. K., N. Koeniger, and G. Koeniger. 1990. Congregation of *Apis cerana indica* Fabricius 1798 drones in the canopy of trees in Sri Lanka. *Apidologie* 21: 201-208.

Rath, W. 1992. The key to Varroa: The drones of *Apis cerana* and their cell cap. *Am. Bee. J.* 132: 329-331.

Rinderer, T. E., B. P. Oldroyd, S. Wongsiri, H. A. Sylvester, L. I. de Guzman, S. Potichot, W. S. Sheppard, and S. L. Buchmann. 1993. Time of drone flight in four honey bee species in south-eastern Thailand. *J. Apic. Res.* 32: 27-33.

Ruttner, F. 1988. Biogeography and Taxonomy of Honeybees. Springer-Verlag, Berlin.

Ruttner, F. and V. Maul. 1983. Experimental analysis of reproductive interspecific isolation of *Apis mellifera* L. and *Apis cerana* Fabr. *Apidologie* 14: 309-327.

Sakagami, S. F. 1960. Preliminary report on the specific difference of behaviour and other ecological characters between European and Japanese honeybees. *Acta Hymenopterologica* 1: 171-198.

Salmah, S., T. Inoue, and S. F. Sakagami. 1990. An analysis of apid bee richness (Apidae) in Central Sumatra. *In:* Sakagami. S. F., R. Ohgushi, and D. W. Roubik (Eds.), Natural History of Social Wasps and Bees in Equatorial Sumatra, pp. 139-174. Hokkaido University Press, Sapporo.

Seeley, T. D. 1985. Honeybee Ecology. Princeton Univ. Press, Princeton, NJ.

Shapiro, A. M. and A. H. Porter. 1989. The lock-and-key hypothesis: evolutionary and biosystematic interpretation of insect genitalia. *Ann. Rev. Entomol.* 34: 231-245.

Simpson, H. 1960. Male genitalia of Apis species. *Nature* 185: 56.

Simpson, H. 1970. The male genitalia of *Apis dorsata* (F.). (Hymenoptera: Apidae). *Proc. R. Entomol. Soc. London A* 45: 169-171.

Smith, D. R. and R. H. Hagen. 1996. The biogeography of *Apis cerana* as revealed by mitochondrial sequence data. *J. Kansas Entomol. Soc. Suppl.* 69: 294-310.

Smith, F. 1861. Description of new species of hymenopterous insects collected by Mr. A. R. Wallace at Celebes. *J. Proc. Linn. Soc. London, Zool.* 5: 57-93.

Tingek, S., M. Mardan, T. E. Rinderer, N. Koeniger, and

G. Koeniger. 1988. Rediscovery of *Apis vechti* (Maa, 1953): the Saban honeybee. *Apidologie* 19: 97-102.

Tingek, S., G. Koeniger and N. Koeniger. 1996. Description of a new cavity dwelling species of *Apis* (*Apis nuluensis*) from Sabah, Borneo with note on its occurrence and reproductive biology (Hymenoptera, Apoidea, Apini). *Biologica Senckenbergiana* 76: 115-119.

Verma, L. R. 1993. Declining genetic diversity of Apis cerana in Hindu Kush-Himalayan region. *In:* Connor, L. J., T. Rinderer, H. A. Sylvester, and S. Wongsiri (Eds.), Asian Apiculture. Proc. 1st. Internat. Conf. Asian Honey Bees and Bee Mites, pp. 81-88. Wicwas Press, Cheshire, CT.

Winston, M. L. 1987. The Biology of the Honey Bee. Harvard University. Press, Cambridge, MA.

Winston, M. L. and K. N. Slessor. 1992. The essence of royalty: honey bee queen pheromone. *Amer. Scientist* 80: 375-386.

Wongsiri, S., K. Limbipichai, P. Tangkanasing, M. Mardan, T. E. Rinderer, H. A. Sylvester, G. Koeniger, and G. Otis. 1990. Evidence of reproductive isolation confirms that *Apis andreniformis* (Smith, 1858) is a separate species from sympatric *Apis florea* (Fabricius, 1787). *Apidologie* 21: 47-52.

Yoshida, T. 1994. Difference in drone congregation areas of native *Apis cerana japonica* Rad. and introduced *A. mellifera L. Jpn. J. Appl. Entomol. Zool.* 38: 85-90.

Yoshida, T., J. Saito, and N. Kajigaya.1994. The mating flight times of native *Apis cerana japonica* Radoszkowski and introduced *Apis mellifera* L. in sympatric conditions. *Apidologie* 25: 353-360.

Zmarlicki, C. and R. A. Morse. 1963. Drone congregation areas. *J. Apic. Res.* 2: 64-66.

Biology of parasitic Asian bee mites and their honey bee hosts

NIKOLAUS KOENIGER
INSTITUT FÜR BIENENKUNDE (POLYTECHNISCHE GESELLSCHAFT)
FACHBEREICH BIOLOGIE DER J.W. GOETHE-UNIVERSITÄT FRANKFURT A.M.
KARL-VON-FRISCH-WEG 2, 61440 OBERURSEL 1, GERMANY

email: bienenkunde@em.uni-frankfurt.de

I. The mites

According to Eickwort (1988) there are 7 species of mites which are considered to be obligatory parasites of honey bees. Since then three species were added (Table 1). *Varroa underwoodi* (Delfinado-Baker and Aggarwal 1987) was number 8; *Euvarroa wongsirii* (Lekprayoon and Tangkanasing 1991) number 9 and the species described most recently (number 10) is *Varroa rindereri* (de Guzman and Delfinado-Baker 1996).

The parasitic honey bee mites belong to two different systematic groups of Acari the Prostigmata and Mesostigmata. The Prostigmata contain the genus of *Acarapis* with three species. The life cycle of these species is rather uniform. The nymphs and adult mites pierce the cuticle of adult honey bees. *Acarapis dorsalis* and *Acarapis externus* are external parasitic forms. Whereas *Acarapis woodi* feeds commonly inside the main thoracic (mesothorax) trachea. *Acarapis dorsalis* and *Acarapis externus* seem to cause no damage of economic significance in bee colonies. Therefore most of the recent research is restricted to *Acarapis woodi* which has apparently detrimental effects on *Apis mellifera* colonies in North America (see Sammataro this volume). All members of the genus *Acarapis* were reported of *Apis mellifera* and a common evolutionary origin of these species on *Apis mellifera* is rather likely. Recent reports of *Acarapis woodi* from *Apis cerana* in Asia seem to be connected to importation of *Apis mellifera* and transfer of *Acarapis woodi*

from its original host to *Apis cerana*.

The second group of parasitic honey bee mites, the Mesostigmata, contain two different families the Laelapidae and Varroidae. The parasitic members of these groups reproduce and complete their larval development on brood stages of the bees. The genus *Tropilaelaps* and the Varroidae (with the two genera, *Varroa* and *Euvarroa*) are of Asian origin.

As of 1997, the Varroidae contain 5 species: *Euvarroa sinhai* Delfinado-Baker 1974, *Euvarroa wongsirii*, *Varroa jacobsoni* Oudemans 1910, *Varroa underwoodi* and *Varroa rindereri*. The genus *Tropilaelaps* consists of the two species *Tropilaelaps clareae* Delfinado and Baker 1961 and *Tropilaelaps koenigerum*. Delfinado-Baker and Baker 1982.

II. Bees and mites

The Western honey bee *Apis mellifera* is naturally allopatric to the other *Apis* species (Ruttner 1988), which are summarized as "Asian" in this paper (though the eastern parts of *Apis mellifera*'s natural distribution cover a large portion of the Asian continent). When *Apis mellifera* is brought into contact with the Asian honey bee species, the natural balance among the sympatric bee species and their parasites is disturbed (Koeniger and Koeniger 1984). Already, two Asian mite species (*Varroa jacobsoni*, Tropilaelaps clareae) have recently changed over to *Apis mellifera*. These 'human-made' associations of mites and bees will be discussed separately and later.

A. Natural associations of mites and bees
The genus *Varroa*

In many areas of Asia several *Apis* species are found in the same habitat. Interspecific robbing and competition for nectar sources occurs frequently (Koeniger 1982). Therefore contacts among honey bees of different species exist which frequently permit parasites to reach another species. Nevertheless data which are available from field studies of *Apis* populations (from areas without exotic *Apis mellifera*) seem to indicate that the parasitic mites show a significant specific distribution (Koeniger et al. 1983).

Varroa jacobsoni generally is found on *Apis cerana*, where it reproduces. Further, we have found few *Varroa jacobsoni* in colonies of *Apis koschevnikovi* in Malaysia. The question whether these mites accidentally reached these colonies from nearby *Apis cerana* could not be decided. In any case there seems to be no report on reproduction of *Varroa jacobsoni* in any other Asian honey bee besides *Apis cerana*.

Varroa underwoodi was reported from *Apis cerana* in India, Korea and several others Asian areas including Java and Borneo. Details of its life cycle and its coexistence with *Varroa jacobsoni* are not yet (1997) available.

Recently *Varroa rindereri* was described from *Apis koschevnikovi* colonies of Borneo (de Guzman and Delfinado-Baker 1996). Since *Varroa rindereri* is very similar to *Varroa jacobsoni*, we can assume that previous reports of *Varroa jacobsoni* in *Apis koschevnikovi* are the result of misidentification because of the close similarity.

Summarizing, the genus *Varroa* seems to be naturally restricted to the cavity dwelling honey bee species of Asia (Table 2). Though sometimes only in a few meters distance of *Apis cerana* colonies infested with *Varroa jacobsoni* colonies of *Apis florea*, *Apis dorsata* or *Apis andreniformis* are found, the free nesting honey bee species do not host *Varroa jacobsoni* (Koeniger et al. 1983).

The genus *Euvarroa*

Euvarroa sinhai is found in colonies of *Apis florea* and according to our observations also in *Apis andreniformis* in Malaysia. A different species *Euvarroa wongsirii* was described from *Apis andreniformis* in Thailand (Lekprayoon and Tangkanasing 1991). The question, however, is still open, whether both mite species of *Euvarroa* are restricted to one honey bee species only or alternatively, *Euvarroa sinhai* and *Euvarroa wongsirii* share *Apis florea* and *Apis andreniformis* as common hosts (Table 3).

The genus *Tropilaelaps*

Tropilaelaps clareae and *Tropilaelaps koenigerum* are naturally limited to *Apis dorsata* and *Apis laboriosa* (Table 4). *Tropilaelaps koenigerum* was discovered in colonies of *Apis dorsata* in Sri Lanka. Very recently we have found also *Tropilaelaps clareae* in Sri Lanka. *Tropilaelaps koenigerum* and *Tropilaelaps clareae*, both mite species were found in the same colonies of *Apis laboriosa* in Nepal. Further, there are new records of *Tropilaelaps koenigerum* from *Apis dorsata* in Thailand. Occasionally *Tropilaelaps clareae* was reported from *Apis cerana*. But apparently these mites can not reproduce or survive for very long.

B. Parasite-host relations
I. *Euvarroa*

According to Akratanakul and Burgett (1976) the life cycle of *Euvarroa sinhai* represents the typical characteristics of the family Varroidae. The adult (mated) females are morphologically well adapted to ectoparasitism on adult bees (drones, workers and queens). For reproduction the *Euvarroa sinhai* female enters drone brood cells which contain old larvae. During the metamorphosis and the development of the drone the mite reproduces. From the eggs the development via deutonymphs and tritonymphs proceeds to the adult mites. The males

Table 1. List of mites obligatory parasitic to *Apis* (1997)			
Prostigmata	**Mesostigmata**		
	Varroidae		**Laelapidae**
AcarApis woodi (1)	Varroa jacobsoni (4)	Euvarroa sinhai (7)	Tropilaelips clareae (9)
AcarApis dorsalis (2)	Varroa underwoodi (5)	Euvarroa wongsiri (8)	Tropilaelaps koenigerum (10)
AcarApis externus (3)	Varroa rindereri (6)		

Table 2. The genus _Varroa_ and its hosts, the cavity dwelling _Apis_ species

	Varroa jacobsoni	_Varroa underwoodi_	_Varroa rindereri_
A. cerana (A. nigrocincta)	yes	yes	yes (?)
A. koschevnikovi	yes (?)	no (?)	yes
A. nuluensis	yes (?)	(yes)	?
A. mellifera	yes	no (?)	?

have reduced chelicers and do not feed. They mate the young females within the brood cell and die after the emergence of the drone. Though we inspected hundreds of worker brood cells of _Apis florea_ no infestation of _Euvarroa sinhai_ was found (Koeniger _et al._ 1983). The numbers of _Euvarroa sinhai_ mites found per colony in Sri Lanka were low. In Thailand and Iran higher degrees of infestation were reported (Mossadegh 1991, Koeniger _et al._ 1993). No observation on the biology of _Euvarroa wongsirii_ in _Apis andreniformis_ are yet available.

2. Varroa

Varroa jacobsoni was discovered 1904 in an _Apis cerana_ colony in Java. But this relatively large mite remained unnoticed for more than 50 years and no further reports were available until a few years ago. The main reason seems to be that _Varroa jacobsoni_ does not cause any apparent damage to the _Apis cerana_ colony. Several mechanisms may contribute to a regulation of _Varroa jacobsoni_ population in an _Apis cerana_ colony:

a. The reproduction of _Varroa jacobsoni_ is restricted to drone brood. The mites do not reproduce in worker brood cells (Koeniger _et al._ 1981). It was thought first that this is a contribution of _Varroa_ to limit its own multiplication in order not to risk the death of its host colony. But Rath and Drescher (1990) recently have found that _Apis cerana_ actively opens worker brood where _Varroa_ reproduces.

b. In Java we observed _Varroa jacobsoni_ in a larger apiary with more than 500 _Apis cerana_ colonies in

one location (Sukabumi). We inspected sealed drone brood cells in 3 colonies and found an infestation rate of 57%. Among the 'normal' cells several 'old' drone cells with dark or black cell capping were noticed. These cells contained many (up to 28/cell) dead _Varroa jacobsoni_ beneath a dead drone pupa (Koeniger _et al._ 1983). Hence, _Apis cerana_ workers did not open sealed drone brood cells after the pupa had died and _Varroa jacobsoni_ were trapped and died.

c. A special grooming behavior of _Apis cerana_ seems to cause a elimination of _Varroa jacobsoni_ from adult bees (Peng _et al._ 1987). Very recently new data were presented which indicate a lower mite elimination than that reported earlier (Fries _et al._ 1996). Life cycle and host adaptations for _Varroa underwoodi_ and _Varroa rindereri_ are still unknown.

3. Tropilaelaps

Woyke (1985) demonstrated that _Tropilaelaps clareae_ feeds on brood mainly and can survive on adult bees of _Apis mellifera_ no longer than 48 hours. Cage tests with adult bees of _Apis mellifera_, _Apis cerana_ and _Apis dorsata_ showed a slightly longer survival of the mites on _Apis dorsata_ worker bees (Koeniger and Muzaffar 1988). Inspections of the dead mites showed that _Apis dorsata_ workers apparently were able to injure _Tropilaelaps clareae_. _Apis dorsata_ actively grooms and hunts for the mites. The mites found under the cages of _Apis mellifera_ were not mutilated.

A second mechanism of mite elimination is connected to colony migration. We inspected several empty combs after the bees had left in the course of absconding in Thailand and Malaysia. In each comb we found a number of capped cells which contained dead or alive pupae with a number _Tropilaelaps clareae_ mites. We inspected them and found large numbers of mites going up to 25 and

Table 3. The genus _Euvarroa_ and its _Apis_ host species

	Euvarroa sinhai	_Euvarroa wongsirii_
A. florea	yes	no
A. andreniformis	yes	yes
A. mellifera	no	no

28 per cell. We calculated that the bees had left more than 1000 *Tropilaelaps clareae* behind in a single comb. Similar observations were made in Thailand (Koeniger and Koeniger 1993).

III. Asian mites and *Apis mellifera*

1. Varroa

Varroa jacobsoni—In *Apis mellifera*, *Varroa jacobsoni* reproduces in worker brood cells. There is a significant preference for drone brood and the reproductive success is higher in drone cells than in worker brood. The 'choice' of *Varroa jacobsoni* between worker and drone brood seems to be influenced by selection. In Europe this modus of reproduction of *Varroa jacobsoni* seems to increase the number of mites in an infested colony until finally the colony will collapse (or the beekeeper applies acaricides).

European *Apis mellifera* in Europe does not posses a natural tolerance to Varroatosis. Comparisons among the European races did not result in significant differences in reproduction (Büchler 1990, Le Conte and Cornuet 1989, Otten 1991). But within each population some variation exists. The attractiveness of brood for *Varroa*, the number of infertile female mites in worker brood, grooming behavior and the duration of the cell capping phase varies among colonies. Several observations indicate differences in the growth of *Varroa* population within an infested colony. But, until now (1997) we are not aware of any real prospect or proof for resistance to *Varroa* on a practical level within Europe.

The question however whether the species *Apis mellifera* is able to resist *Varroa jacobsoni* can be answered positively. In some regions of South America and North Africa large populations of *Apis mellifera* live as part of nature without the intensive care of beekeepers. Though being brought into contact with *Varroa jacobsoni* not earlier than the honey bees in Europe these African and South American *Apis mellifera* populations have already acquired some ways to deal successfully with *Varroa jacobsoni* (Ruttner *et al.* 1984, Ritter 1988, Moretto *et al.* 1991).

The 'other' *Varroa* species — *Varroa underwoodi* is reported from Korea and other areas of Asia into which *Apis mellifera* is kept by humans and up to now (1997) no infestation of *Varroa underwoodi* of *Apis mellifera* has been reported. *Varroa rindereri* was found in Borneo (Malaysia), in a region where indigenous honey bees are kept and where no *Apis mellifera* were imported. So, we can assume that there was not yet any possibility of contact between *Varroa rindereri* and *Apis mellifera*. Therefore the potential of *Varroa rindereri* parasitizing *Apis mellifera* colonies is unknown. In any case both species *Varroa underwoodi* and *Varroa rindereri* should be considered to be a mayor threat to *Apis mellifera*.

2. Tropilaelaps

In subtropical and tropical Asia the 'main' damage of imported *Apis mellifera* results from infestations of *Tropilaelaps clareae*. Frequently *Tropilaelaps clareae* has apparently a higher rate of reproduction (Ritter and Ritter-Schneider 1988) and outnumbers *Varroa jacobsoni* which often occurs in the same *Apis mellifera* colonies. Survival of *Tropilaelaps clareae* depends on the presence of honey bee brood in the colony (Woyke 1985).

The life cycle and the biology of *Tropilaelaps koenigerum* is still unknown. Both species however are morphological very similar. It is difficult to distinguish between *Tropilaelaps clareae* and *Tropilaelaps koenigerum* without proper equipment (microscope etc.). So, until other evidence is available *Tropilaelaps koenigerum* must be considered at least as detrimental to *Apis mellifera* as *Tropilaelaps clareae*.

3. Euvarroa

The question of whether *Euvarroa sinhai* has the potential to successfully parasitize *Apis mellifera* is of interest. Naturally *Euvarroa sinhai* is found in strict association with *Apis florea* (Koeniger and Koeniger 1984). In Southern Iran the distribution of *Apis florea* and *Apis mellifera* overlap (Ruttner 1988). Mossadegh (pers. communication) however

Table 4. The genus *Tropilaelaps* and its *Apis* host species

	Tropilaelaps clareae	*Tropilaelaps koenigerum*
A. dorsata	yes	yes
A. laboriosa	yes	yes
A. mellifera	yes (Asia only ?)	?

did not find *Euvarroa sinhai* in *Apis mellifera* colonies. Two reported findings of *Euvarroa sinhai* occurred in India and Northern Thailand (Kapil and Aggarwal 1987, Rath and Delfinado-Baker 1990). The dead mites were discovered in the hive debris of imported *Apis mellifera* colonies.

Mossadegh (1990) demonstrated that *Euvarroa sinhai* can develop on worker and drone brood of *Apis mellifera* under laboratory conditions. Koeniger *et al.* (1993) tested *Euvarroa sinhai* in cages with worker bees of *Apis mellifera*. *Euvarroa sinhai* mites were able to infest and feed on *Apis mellifera* worker bees. The success rate of *Euvarroa sinhai* was only slightly lower on *Apis mellifera* (53% survival) compared to *Euvarroa sinhai* on worker of *Apis florea* (70% survival). More experiments are needed to further explore the potential of *Euvarroa sinhai* to parasitize *Apis mellifera*.

IV. Phylogeny of parasite-host relationship of Varroidae

The Varroidae are found in the two taxonomic most distant honey bee groups, the small free nesting species (*Apis andreniformis*, *Apis florea*) and the cavity dwelling species. Therefore it seems probable that the genus *Apis* and the Varroidae as well are monophyletic groups. Hence, the common ancestor of *Apis* must have had a parasitic ancestral *Varroa*-type mite. From this origin *Apis cerana* with *Varroa jacobsoni* on one side and on the other side *Apis florea* with *Euvarroa sinhai* may have evolved.

The life cycle of *Varroa jacobsoni* and *Euvarroa sinhai* shows many common characteristics. In both mite species the adult females are found on drones and workers. For reproduction they enter into brood cells. So, the functional division of the life cycle into two parts can be regarded as an old character.

The degree of drone brood preference differs significantly between both species. *Varroa jacobsoni* has a preference for drone brood, but "still" *Varroa jacobsoni* enters worker brood of *Apis cerana*. Normally, the mites do not start oviposition on worker larvae. They stay without offspring in the sealed brood cell until the young bee opens it and emerges. Then, the mite is set free again and will move on to a bee or to another brood cell again. In *Euvarroa sinhai*, by contrast the mites enter only into drone brood. They are not found in worker brood at all.

Apparently evolution went over three stages:

1. A mite reproducing on worker and drone brood causes damage to the host colony (like *Varroa jacobsoni* and *A. mellifera*).

2. Bees which "recognize" reproduction of *Varroa* in worker cells open these brood cells and eliminate the *Varroa* types which reproduce on worker brood. This "hygienic" behavior is common in *Apis cerana*.

3. The mite enters both types of brood but reproduction is limited to drone brood (like *Varroa jacobsoni* and *Apis cerana*). During the time which the mite spends in a worker brood cell (without reproduction), it might miss a drone brood cell which allows reproduction.

4. In the final stage the female mite avoids worker brood and enters in drone cells only. According to this hypothesis the *Euvarroa sinhai/Apis florea* relation has reached the most advanced stage.

Literature

Akratanakul P., M. Burgett (1976) *Euvarroa sinhai* Delfinado and Baker (Acarina: Mesostigmata): a parasitic mite of *Apis florea*. *J.Apic.Res*.15, 11-13

Büchler R. (1990) Möglichkeiten zur Selektion auf erhöhte Varroa -Toleranz mitteleuropäischer Bienenherkünfte. *Apidologie* 21, 365-367

Le Conte Y., Cornuet J.M. (1989) Variability of the post-capping stage duration of the worker brood in three different races of *Apis mellifera*.In: Present status of Varroatosis in Europe and progress in the Varroa mite control. Proc meeting EC expert's group. Ed. Cavalloro R, pp 171-174

Delfinado M., (1963) Mites of the honeybee in South-East Asia. *J.Apic.Res* . 2 (2), 113-114

Delfinado M., Aggarwal K. (1987) A new species of *Varroa* (Acari, Varroidae) from the nest of *Apis cerana* (Apidae). *Int. J.Acarol.* 13, 233-237

Delfinado M., Baker E.W. (1961) *Tropilaelaps* a new genus of mite from the Philippines (Laelaptidae s.lat.: Acarina) *Fieldiana Zool* 44, 53-65

Delfinado-Baker M., Baker E.W. (1982) A new species of *Tropilaelaps parasitic* on honey bees *Am. Bee J.* 122, 416-417

de Guzman L.I., Delfinado-Baker M. (1996) A new species of *Varroa* (Acari, Varroidae) associated with *Apis koschevnikovi* (Apidae, Hymenoptera). *Int. J.Acarol.* 22, 23-27

Eickworth G.C. (1988) The origins of mites associated with honey bees *In:* Africanized honey bees and bee mites. Eds Needham G.R., Page E.R., Delfinado-Baker M., Bowman C.E. Ellis Horwood Series in Entomology and Acaraology. John Wiley & Sons, New York, pp 327-338

Fries I, Wei Huazhen, Shi Wie, Chen Su Jin (1996) Grooming behavior and damaged mites *(Varroa jacobsoni)* in *Apis cerana cerana* and *Apis mellifera ligustica*. *Apidologie* 27, 3-11

Kapil R.P., Aggarwal K. (1987) *Euvarroa sinhai* found in *Apis mellifera* hive debris. *Bee World* 68, 189

Koeniger N. (1982): Interactions among the four species of the genus *Apis*. *In:* The biology of social insects. Eds Breed M.D., Michener C.D., Evans H.E., Westview Press, Boulder, USA

Koeniger N. (1990) Coevolution of Asian honey bees and their parasitic mites. Proc. 11th Int Congr. IUSSI, India pp. 130-131

Koeniger N., Koeniger G., Delfinado-Baker M. (1983) Observation on mites of the Asian honeybee species (*Apis cerana, Apis dorsata, Apis florea*). *Apidologie* 14 (3), 197-204

Koeniger N., Fuchs S. (1988): Control of *Varroa jacobsoni*: current status and developments. *In:* Africanized honey bees and bee mites. Eds Needham G.R., Page E.R., Delfinado-Baker M., Bowman C.E. Ellis Horwood Series in Entomology and Acaraology. John Wiley & Sons, New York, pp 360-369

Koeniger N., Koeniger G., (1984) Change of host by parasitic mites in Asia after a new honeybee species is introduced. Proc 3rd Int Conf Apic Trop Climates, Nairobi, pp 160-162

Koeniger N., Koeniger G. (1993) Possible effects of regular treatments of Varroatosis on the host-parasite relationship between *Apis mellifera* and *Varroa jacobsoni In:* Asian Apiculture. Proceedings of the 1st Int. Conference on the Asian Honey Bees and Bee Mites. Bangkok. L.J. Connor *et al.*, editors) 1992, pp. 541-550

Koeniger N., Koeniger G., de Guzman L.I., Lekprayoon C. (1993) Survival of *Euvarroa sinhai* Delfinado and Baker (Acari, Varroidae) on workers of *Apis cerana* Fabr, *Apis florea* Fabr and *Apis mellifera* L in cages. *Apidologie* 24: 403-410.

Koeniger N., Koeniger G., Wijayagunesekara H.N.P. (1981) Beobachtungen über die Anpassung von *Varroa jacobsoni* an ihren natürlichen Wirt *Apis cerana* in Sri Lanka. *Apidologie* 12 (1), 37-40

Koeniger N, Muzaffar N (1988) Lifespan of the parasitic honeybee mite, Tropilaelaps clareae, on *Apis cerana, dorsata and mellifera. J. Apic. Res.* 27 (4), 207-212

Lekprayoon C., Tangkanasing P. (1991) *Euvarroa wongsiri*, a new species of bee mite from Thailand. *Int. J .Acarol.* 17, 225-258

Moretto G., Goncalves L.S., de Jong D., Bichuette M.Z. (1991) The effects of climate and bee race on *Varroa jacobsoni* infestations in Brazil. *Apidologie* 22, 197-203

Mossadegh M.S. (1990) Development of Euvarroa sinhai (Acarina: Mesostigmata), a parasitic mite of *Apis florea*, on *Apis mellifera* worker brood. *Exp. Appl. Acarol.*, 73-78

Mossadegh M.S. (1991) Geographical distribution and levels of infestations of *Euvarroa sinhai* Delfinado and Baker (Acarina: Mesostigmata), in *Apis florea* colonies in Iran. *Apidologie* 22, 127-134

Otten C. (1991) Vergleichende Untersuchungen zum Populationswachstum von *Varroa jacobsoni* in Völkern von *Apis mellifera* unterschiedlicher geographischer Herkunft. Dissertation, Institut für Bienenkunde, Fachbereich Biologie, J.W. Goethe-Universität, Frankfurt a. M., Germany

Peng Y.S., Fang Y., Xu S., Ge L. (1987): The resistance mechanism of the Asian honeybee *Apis cerana* to an ectoparasitic mite *Varroa jacobsoni. J .Invertebr. Pathol* .49, 54-60

Rath W., Delfinado-Baker M. (1990) Analysis of *Tropilaelaps clareae* populations from the debris of *Apis dorsata* and *Apis mellifera* in Thailand. *In:* Proc Int Symp Recent Res on Bee Pathology (Ritter W, ed) Ghent, 86-89

Rath W., Drescher W. (1990) Response of *Apis cerana* towards brood infested with *Varroa jacobsoni* and infestation rate of colonies in Thailand. *Apidologie* 21, 311-321

Ritter W. (1988) *Varroa jacobsoni* in Europe, the tropics, and subtropics *In:* Africanized honey bees and bee mites. Eds Needham G.R., Page E.R., Delfinado-Baker M., Bowman C.E. Ellis Horwood Series in Entomology and Acaraology. John Wiley & Sons, New York, pp 349-359

Ritter W., Ritter-Schneider U. (1988) Differences in biology and means of controlling *Varroa jacobsoni* and *Tropilaelaps clareae*, two novel parasitic mites of *Apis mellifera. In:* Africanized honey bees and bee mites. Eds Needham G.R., Page E.R,. Delfinado-Baker M., Bowman C.E. Ellis Horwood Series in Entomology and Acaraology. John Wiley & Sons, New York, pp 387-395

Ruttner, F. (1988) Biogeography and taxonomy of honeybees. 284, Springer Verlag, Berlin

Ruttner F., Ritter W. (1980) Das Eindringen von *Varroa jacobsoni* nach Europa im Rückblick. *Allgemeine Deutsche Imkerzeitung* 14 (5), 130-134

Ruttner F., Hänel H. (1992) Active defense against *Varroa* mites in a Carniolan strain of honeybee (*Apis mellifera carnica*). *Apidologie* 23 (2), 173-187

Ruttner F., Marx H., Marx G. (1984) Beobachtungen über eine mögliche Anpassung von *Varroa jacobsoni* an *Apis mellifera* in Uruguay. *Apidologie* 15 (1), 43-62

Woyke J. (1985): *Tropilaelaps clareae* in Afghanistan, and control methods applicable in tropical Asia. Proc 3rd Int Conf Apic Trop Climates Nairobi, pp 163-166

Current status and problems in chemotherapy of Varroatosis

NORBERTO MILANI
DIPARTIMENTO DI BIOLOGIA APPLICATA ALLA DIFESA DELLE PIANTE
UNIVERSITÀ DI UDINE
VIA DELLE SCIENZE, 208, I 33100 UDINE
ITALY

Summary

Effective acaricides for the control of *V. jacobsoni* were developed mainly between 1980 and 1986; of special interest are products with a prolonged action, based both on synthetic acaricides and on substances of natural origin. There is increasing concern regarding the accumulation of residues in beeswax and the possible contamination of honey. An even more serious problem is the selection of resistant varroa strains; the spread of mites resistant to fluvalinate in Italy and more recently to other European countries caused heavy damage to beekeeping. Early detection of the resistance is crucial to reduce damage. Increased tolerance to other active ingredients used in the control of the varroa mite is documented as well. Regarding the resistance management tactics, "moderation tactics" seem to be more appropriate than "high dose tactics" to slow down the selection of resistant varroa strains.

Acaricides currently used against the varroa mite

The spread of the mite *Varroa jacobsoni* Oudemans since the 1960s (Matheson 1995) has posed a major threat to apiculture worldwide and required regular treatments to save the bees. Countless synthetic acaricides, plant extracts and the most diverse empirical preparations have been proposed and tested for the varroa control (cf. Wienands 1988; Milani 1993), but only a small number has provided suitable mite control.

The acaricides introduced into use in the 1960s (chlorobenzilate and phenotiazine) were completely abandoned during the 1980s. Bromopropylate, introduced around 1975, was also superseded due to its difficulty of use and to the high load of residues. Effective acaricides became available during the 1980s, when the mite invaded western Europe.

At present, chemotherapy of varroatosis is based both on synthetic acaricides and on substances of natural origin. Among the former, it is based essentially on four active ingredients introduced into use between 1980 and 1986 (Table I) and on a few closely related substances. During the last decade, no additional unrelated chemicals suitable for varroa control were found and successfully introduced into use. Among the substances of natural origin, formic acid and essential oils (mainly thymol) have also been widely utilized (Table I) since the end of the 1970s. Recently, oxalic acid has been rediscovered and considered with increasing favour by Italian beekeepers. Although the literature on the control of the varroa mite by using natural products is very rich, much less effort has been made to study the toxicity of these substances to the varroa mites and honey bees and other basic parameters needed to optimize their use (cf. Charrière *et al.* 1992). For example, it has only recently been recognized that eucalyptol (which is a component of commercial formulations) is quite toxic to bees at therapeutic concentrations, while other, yet unused compounds seem more suitable (Imdorf *et al.* 1995).

The introduction of the Apistan strips and of analogous products with a prolonged action represented a major progress in the chemotherapy of the varroatosis, making the broodless condition—often incompatible with constraints on the minimum temperature needed for treatments—unnecessary and enabling healthy winter bees to be obtained (most winter bees infested during the pupal stage cannot overwinter: Kovac and Crailsheim 1988). Fluvalinate and other closely related pyrethroids have an exceptionally low toxicity to the honey bee and this reduces the risk of chronic poisoning. Attempts have been made to develop strips with a long duration of action based on other active ingredients, such as amitraz and various organophosphates including bromphenvinphos (Jelionsski 1994; Cieoslak, pers. comm.) and coumaphos (Milani and Iob, unpublished data). The results—only partly made known—seem to indicate that at least some of these products can have a satisfactory efficacy, though the safety margin is much narrower than with fluvalinate; an additional problem is the instability of the amitraz molecule

in the strip formulation.

Long duration treatments with evaporating substances of natural origin have also been developed; *e.g.*, the Krämer Platte (Wachendörfer *et al.* 1983) and its modifications, which release formic acid over a long period of time and the preparation Var, based on ethereal oils (Rickli *et al.* 1991). Of special interest are some new preparations made of gels, that release the evaporating ingredient at a constant rate, irrespective of the temperature and could represent a major innovation in the chemotherapy of *V. jacobsoni.*

In contrast, the systemic mode of action of varroacide substances (such as coumaphos and cymiazole) has lost part of its interest and should be reconsidered at least for some products. The main supposed advantage of this mode of action, *i.e.* the lack of contamination of the hive products, does not take place (in an experiment, only one fourth of the applied coumaphos was taken up by the bees: van Buren *et al.* 1992) and it is possible that some of these substances act by direct contact instead. Moreover, the formulation seems to be rather inefficient: though coumaphos is about 6 times more toxic to varroa than fluvalinate (Milani

---

Table 1. Main acaricide substances presently used in the chemotherapy of varroatosis. The name in italics indicate the substances which are closely related to the preceding one.

SUBSTANCE	GROUP	YEAR INTRODUCED	REFERENCES
SYNTHETIC ACARICIDES			
Coumaphos	Organophosphate	1980	Ho and An 1980
a.i. of the Cekafix	id.	1991	Schabitz et al. 1991
Amitraz	Formamidine	1981	Csaba and Kavai 1981; Kilani et al.; 1981
Cymiazole	Thiazoline	1985	Ritter 1985
τ-Fluvalinate	Pyrethroid	1986	Borneck 1986
Flumethrin	id.	1986	Koeniger 1986
Acrinathrin	id.	1993	Vesely 1993
NATURAL SUBSTANCES			
Thymol	Ethereal oil	1978	Mikityuk and Grobov 1978
Formic acid	Organic acid	1979	Kunzler et al. 1979
Lactic acid	id.	1983	Koeniger et al. 1983
Oxalic acid	id.	1983	Popov 1983, Koeniger

1995; Milani and Della Vedova 1996), the dose applied in a treatment with Perizin is about 10 times higher than that needed to obtain similar results with fluvalinate in short duration treatments (Senegaocnik 1991).

The availability of acaricide products which were very effective against *V. jacobsoni*, tolerated by the honey bee and easy to use made the control of the mite much easier during the late 1980s, and the idea that varroatosis had been almost overcome as a problem began to gain ground at the beginning of the 1990s. Although aware of the limits of chemotherapy—the buildup of residues and the selection of resistant varroa strains—we could better appreciate their impact during the last years.

Residues of acaricides in hive products

During the early 1980s, when the varroacide products were being developed and tested, there was more concern about their effectiveness and tolerance by the honey bee than about possible residues.

However, it was soon recognized that pyrethroids (Taccheo Barbina and De Paoli 1993) are extremely stable in honey and that coumaphos and bromopropylate also show a remarkable stability (Taccheo Barbina *et al.* 1988, 1989); thus the residue levels depend essentially on the amount applied and on the dilution by the imported honey. Individual samples of honey differ widely in their residue content, depending on the history of treatments of the hive. The contamination of the honey after a treatment with bromopropylate (typically, 50-150 ppb) is in general higher than that of coumaphos (5-30 ppb), while the residues of fluvalinate only exceptionally exceed 10 ppb (usually as a result of illegal application (Cardenal Galván *et al.* 1989). The extremely small amounts of flumethrin used in treatments and its scarce solubility in water make the residues of this acaricide undetectable.

The above mentioned varroacides are highly lipophilic, thus they build up in the wax and in the propolis where they reach high levels (even 100 ppm or more) as a result of repeated applications (Wallner 1995, 1996, de Greef *et al.* 1994); they can be eventually released into the honey (Wallner 1995) or contaminate it indirectly, being contained in fine wax particles that are difficult to separate from the honey. Moreover, they are often present in the wax foundations (Wallner 1996) and thus can diffuse

into the honey even if the beekeeper never used them. Direct contact between pyrethroid containing strips and adjacent combs causes an especially high accumulation of residues in the wax of the latter (Wallner *et al.* 1995); unfortunately, the strips were not designed to avoid this contact. Cleaning the wax from the residues seems to be difficult or even impossible; the treatments which can reduce the contamination also affect the chemical composition of the wax (Vesely *et al.* 1994; Wallner 1996) and the development of a technique to hydrolize the fluvalinate residues in the wax has apparently been discontinued.

Amitraz and cymiazole hydrochloride are less lipophilic and can be found in the honey; the residues of the former can reach 100 ppb immediately after treatment, but decay with a time constant depending on the pH of the honey and varying between a few days and two months (Bogdanov 1988; Franchi and Severi 1989); unfortunately, the metabolites are much more stable (Franchi and Severi 1989). The residues of cymiazole can reach 2000 ppb after treatment, but are stable in the honey and the decrease to 200 ppb observed occurs later because of the dilution of the a.i. (Cabras *et al.* 1994).

Thymol and formic acid often leave comparatively higher levels of residues, since the amount applied is much larger; they may affect the organoleptic qualities of the honey but tend to escape by evaporation in a few months. Thymol can reach 178 ppm in the wax and 3 ppm in the honey in the spring following an autumn treatment (Lodesani *et al.* 1992). Formic acid is found only in the honey, in concentrations up to 1000-2000 ppm immediately after a treatment, but decreasing to 50-100 ppm after four months (Stoya *et al.* 1986; Hansen and Guldborg 1988); the final levels are lower than those naturally occurring in some honeydew or chestnut honeys (up to 600 ppm). Little is known about the residues of oxalic acid, but they may not be negligible due to the high amount of this compound used in each treatment (up to 10 g/colony).

The synthetic products used to control varroa do not have bad toxicological profiles, except amitraz, whose main metabolite (2,4-dimethyl-aniline) is a suspected carcinogen. Little is known about possible long term toxicity of the products of natural origin, which are "generally regarded as safe" and are present in many foodstuffs. However, a Russian study found that formic acid is mutagenic at high doses (Alekseenko *et al.* 1985), but more data are needed to reach a reliable conclusion. In every case, the exposure and the risk seem to be much higher for the beekeeper than for the consumer; lethal risks could be incurred in the manipulation of the oxalic acid, which is extremely dangerous if inhaled.

Our knowledge of the residues of co-formulants, impurities and breakdown products present in the acaricides used for the treatments is still inadequate. A warning was issued about epichlorhydrine, a suspected carcinogen used as a stabilizer of amitraz (Schieferstein 1985). A study of the aerosol produced with a fogger revealed the presence of a wide spectrum of compounds, besides the active ingredient amitraz (Faucon *et al.* 1986).

Especially in some countries, the presence of residues of synthetic acaricides are often felt to be a major problem in the chemotherapy of varroatosis and beekeepers are encouraged to use only products of natural origin or non chemical control techniques. Unfortunately, these had little success in areas with a warm climate. Studies aimed at giving realistic suggestions on how to reduce the residues and still effectively control the varroa mite are scarce. The contamination of hive products depends essentially on the amount of acaricide applied in treatments and thus could be reduced—without placing the bees at a risk—by decreasing the duration of the treatment with products formulated in strips (an efficacy exceeding 99% is not always really needed and can be rapidly twarted by reinfestation) or improving the "systemic" formulations.

Selection of resistant mites

A second, more serious limit of chemotherapy is the selection of resistant mites. The onset of resistance is a worrying problem in crop protection and is usually overcome by changing the a.i. used. In the case of varroa, the problem is made critical by the scarce number of unrelated chemicals suitable for control.

The existence of strains of *V. jacobsoni* with an increased tolerance to the active ingredients used is described for most of the widely used acaricides.

With regard to bromopropylate, Ritter and Roth (1988) presented a report at a European Meeting showing a gradual increase in the LD_{50} of mites taken from nuclei repeatedly treated with low doses of that a.i..

The inefficacy of amitraz in Vojvodina, a region of the former Yugoslavia, after four years of use was clearly document by Dujin *et al.* (1991) and colony losses were recorded in the nearby Hungary. It was rumoured that the efficacy of amitraz was unsatisfactory in some regions of Italy and France (Celli and Porrini 1987). An indirect indication of the existence of mixed susceptible and resistant populations could be given by a dose-mortality curve extending over three orders of magnitude found in laboratory tests using a topical application technique (Abed and Ducos de Lahitte 1993); unfortunately, only one population was studied. Little attention was paid to this problem later since amitraz was substituted with fluvalinate.

The resistance to fluvalinate in Italy had a much more serious impact on beekeeping and was better studied; it seems, however, that Apistan treatments were not fully satisfactory also in a region of Egypt (Abou-Zaid and Ghoniemy 1992) and possibly in other Mediterranean countries. In 1991-92 reduced efficacy of Apistan was reported from Lombardy (Loglio and Plebani 1992; Colombo *et al.* 1993), a region where the veterinary services had given free Apistan strips to all the beekeepers. Similar problems seem to have occurred in Sicily, but precise data are lacking; although these areas are widely apart, they are connected by active migratory beekeeping.

The consequences were disastrous bee losses, over 70% in some districts where accurate statistics are available, but locally exceeding 90%, usually unexpected since no abnormal increase in the varroa infestation or unsatisfactory efficacy of treatments had been observed during the previous years. Although aware of the risk of resistance, we were not prepared for it. The situation was made even worse by the fact that the manufacturer of Apistan initially denied the presence of resistance on the basis of an existing laboratory assay (Watkins 1996) and thus appropriate measures were not taken immediately. The resistant varroa mites spread very quickly, due to the high selection pressure (nearly all the colonies, except feral swarms, over all the country were treated with fluvalinate and thus there was very little or no mixing with mites from colonies treated with different active ingredients). Although the initially patchy distribution of the resistant mites can give the impression of parallel, independent selection processes in different regions, all the available data indicate that the resis-

Milani: Problems in chemotherapy of Varroatosis

tance appeared only once and then was spread to all regions of Italy by migratory beekeeping and bee trade. More recently, the fluvalinate resistant strain crossed the Alps—both a geographical and political barrier—and in 1995 and 1996 was detected in nearby countries.

The selection pressure of regular treatments with Apistan strips was evidently strong enough to select resistant mites, and special circumstances do not need to be invoked; however, it still remains difficult to explain the large size of the resistant populations as early as in 1991, after perhaps four years of use of fluvalinate, since mutations giving rise to resistance seem to be extremely rare and in this case associated with a lower fitness.

To study the resistance, we developed a laboratory assay (Milani 1994, 1995) which clearly discriminates susceptible and resistant populations. By using the bioassay, the resistant strain was detected in several localities of France, Switzerland, Austria and Slovenia before any damage was reported (Trouiller, in prep.).

The bioassay was also used to show that the mites surviving Apistan treatments have an increased tolerance to acrinathrin and flumethrin, i.e., that there is cross resistance between these closely related pyrethroids (Milani 1995).

The monooxygenases of the P_{450} system (P_{450} oxidative enzymes are all of a defined molecular weight isolated by ntrifugal density) are at least in part responsible for the detoxication in the varroa mite—a mechanism similar to that of the bee—but other factors could be involved (Hillesheim *et al.* 1996) and the resistance could be based on a polygenic mechanism. A fascinating hypothesis is that the resistant strain integrated into its genome the genes responsible for detoxification in the honey bee (gene transfer from the host insect to entomophagous mites feeding on it has already been observed: Houck *et al.* 1991).

Resistance in insects or mites is often associated with lower fitness (cf. Denholm and Rowland 1992) and this leads to the decline in the frequency of the alleles for resistance (reversion). In the case of the varroa mite resistant to fluvalinate, preliminary data (Greatti and Trouiller, pers. comm.) indicate that increase of the population over one year is in the order of the half of that of the susceptible strain, when kept in hives without any residue of fluvalinate. Thus the reversion does take place but rather slowly.

Early detection of the resistant strains is crucial to reduce the damage. To achieve this goal, in some districts of Italy the Apistan efficacy was checked in the field by carrying out a further treatment with Perizin on a small sample of colonies (1-2% of the total). Although these surveys have some disadvantages (the results often come too late to plan the varroa control and can be influenced by the reinfestation), they proved to be very useful to reduce bee losses. We carried out such a survey in the Pordenone province (northern Italy), in collaboration with the local beekeeping organization (Milani 1996). In 1993 and 1994 the efficacy of the Apistan treatment exceeded 99%, except for two apiaries where the Perizin treatment was delayed and thus reinfestation could not be excluded. In 1995 the efficacy dropped to less than 80% in more than half of the tested apiaries and only in two apiaries was over 99%: the resistant mites had spread almost everywhere. Dangerous levels of resistant mites built up within one year; their presence could hardly be noted without a further treatment. There was a large variability in the percentage of mites surviving Apistan treatments, both within and among apiaries: the resistant mites were often concentrated in a few colonies, which—without further treatments—would have collapsed in the following year and cause a chain of reinfestations and further collapses with disastrous consequences for beekeeping of that region; on the other hand, the variability makes it difficult to extrapolate the information on the proportion of resistant mites to nearby, unchecked apiaries.

Since the spread of strains resistant to fluvalinate, coumaphos was widely used in Italy and in nearby countries. This raised much concern about the possible onset of resistance to this acaricide, especially when a field test showed a decreased efficacy of Perizin in an area of northern Italy. Bioassays carried out in our laboratory (Milani and Della Vedova 1996; Della Vedova *et al.* 1997) showed a slight, 2-3 fold, but constant and significant increase in the tolerance to this acaricide in a sample of mites taken from that area.

Strategies to minimize the risk of resistance

In general, there are two opposite strategies in resistance management (Denholm and Rowlands 1992): the "high dose" strategies aim at eliminating most resistant alleles by using a dose high enough to kill the heterozygotes, supposed endowed with a lower level of resistance (initially, when the frequency of alleles for resistance are extremely low, resistant homozygotes are extremely rare and will be diluted by susceptible individuals immigrating from untreated areas), or by using mixtures of acaricides; the "moderation" strategies aim at reducing the selection pressure, to preserve susceptible individuals, while maintaining the control of the pest.

In the case of the varroa mite—and probably of other mites as well—the assumption that resistant homozygotes are much rarer than heterozygotes is not true because of inbreeding (brother-sister mating), and thus the main stay of the high-dose strategies falls; however, these strategies can be valid if mutations giving a higher degree of resistance are less common or if the resistance is a polygenic character. Redundant killing, by using mixtures of two acaricides, is impossible because of the scarcity of unrelated and "unresisted" active ingredients that can be combined in the same product.

The moderation strategies are probably more appropriate in the case of the varroa mite: the reduction of fitness usually associated with resistance partially balances the selection pressure of the treatments, if the latter is not too high. The chemotherapy of varroatosis should not rely upon a 99% efficacy obtained by the repeated or prolonged use of a single acaricide; a satisfactory control of the parasite should rather be achieved by combining different acaricide treatments, each one acting for a restricted period of time and thus with a reduced effectiveness, or integrating them with non chemical control techniques. Finally, the selection of bee strains on which the increase of the varroa populations is slower would make very highly effective treatments unnecessary and thus would also decrease the risk of selecting resistant mites.

Perspectives

Today, the control of *V. jacobsoni* on European bee races without the use of any acaricide, difficult and time consuming under the more favorable conditions, is impossible in large apiaries or in areas with a Mediterranean climate; the selection of completely tolerant bee strains remains a long term goal, and thus the control of the varroa mite will probably rely on acaricide treatments for several years. We need to develop and to put into practice strategies to reduce the undesirable consequences of the

Milani: Problems in chemotherapy of Varroatosis

treatments; a reduced confidence in a single chemical treatment is part of the strategy to avoid the risk of resistance and could be integrated with the efforts to reduce accumulation of residues in the hive products.

Literature

Abed, T. and J. Ducos de Lahitte. 1993. Détermination de la DL50 de l'amitraze et du coumaphos sur *Varroa jacobsoni* Oud au moyen des acaricides Anti-varroa (Schering) et Perizin (Bayer). *Apidologie* 24: 121-128.

Abou-Zaid, M.I. and H.A. Ghoniemy. 1992. Evaluation of the role of some chemical compounds for controlling *Varroa jacobsoni* Oudemans in Egypt. Magallat Al-Minufiyyat Li-L-Buhut Al-Zira'iyyat (=*Minufiya Journal of Agricultural Research*) 17: 1465-1470.

Alekseenko, A.Ya., G.I. Pavlenko, N.M. Bocharov and E.T. Popov. 1985. Mutagenic activity of curative preparations. *Pchelovodstvo* No. 7, 19-20.

Bogdanov, S. 1988. Bestimmung von Amitraz und seine Metaboliten in Honig durch HPLC. Mitt. Sektion Bienen, Forschungsanstalt fur Milchwirtschaft, Liebefeld, 9 pp.

Borneck, R. 1986. Fluvalinate an interesting molecule in the battle against *Varroa* mites. Proc. Apimondia Int. Symp. Health protection of honey bees, Zagreb 1986.

Cabras, P., M.G. Martini, I. Floris, and L. Spanneda. 1994. Residues of cymiazole in honey and honey bees. *J. Apic. Res.* 33: 83-86.

Cardenal Galván, J.A., A. Gómez Pajuelo and E.C. López-Sepúlveda Garcia. 1989. Using Fluvalinate inserts in Varroa control. Present status of varroatosis in Europe and progress in the varroa mite control, Proc. Meet. EC-Experts' Group, Udine 1988, R. Cavalloro ed., C.E.C., Luxembourg, 339-342.

Celli, G. and C. Porrini. 1987. Amitraz, finalmente dosati i residui. *Ecco un'indagine tutta italiana. Apitalia* 14, no. 3-4: 5-6.

Charrière, J.-D., A. Imdorf and V. Kilchenmann. 1992. Ameisensäure-Konzentrationen in der Stockluft von Bienenvölkern. *Schweiz. Bienenztg.* 115: 463-469.

Colombo, M., M. Lodesani and M. Spreafico. 1993. Resistenza di *Varroa jacobsoni* (Oud) a fluvalinate. Primi risultati di indagini condotte in Lombardia. *L'ape nostra amica* 15: no. 5, 12-15.

Csaba, G. A., and A. Kávai. 1981. Étude sur l'emploi de l'amitraz dans la lutte contre la varroase. Proc. XXVIIIth Intern. Congr. Apicult., Acapulco 1981, Apimondia Publ. House, Bucharest: 307.

de Greef, M., L. de Wael and O. van Laere. 1994. Determination of the fluvalinate residues in the Belgian honey and beeswax. *Apiacta* 29: 83-87.

Della Vedova, G., M. Lodesani and N. Milani. 1997. Resistenza di *Varroa jacobsoni* al coumaphos? *L'ape nostra amica*, 19, no. 1, 6-10.

Denholm, I., and M.W. Rowland. 1992. Tactics for managing pesticide resistance in arthropods: theory and practice. *Annu. Rev. Entomol.* 37: 91-112

Dujin, T., Jovanovic V., Suvakov D. and Milkovic Z. 1991. [Effects of extended utilisation of amytrase-based preparations on the formation of resistant strains of *Varroa jacobsoni*]. *Vet. Glas.* 45: 851-855. In Croatian,

Serbian spelling.

Faucon, J.P., C. Flèche-Séban, C. Flamini, C. Sarrazin, C. Pozzo di Borgo, J. Chevallier, J. Ceccaldi and Y. Chimento. 1986. Le traitement de la varroatose de l'abeille. Évaluation des diverses utilisations de la molécule d'amitraze. *Santé Abeille.* 94, 141-171.

Franchi, A. and A. Severi. 1989. Comportamento dell'amitraz e dei suoi prodotti di degradazione nel miele. Atti VII Simp. Chimica degli Antiparassitari, Piacenza 1989, Biagini, Lucca. 89-96.

Hansen, H. and M. Guldborg. 1988. Residues in honey and wax after treatment of bee colonies with formic acid. *Tidsskr. Planteavl* 92: 7-10.

Hillesheim, E., W. Ritter and D. Bassand. 1996. First data on resistance mechanisms of *Varroa jacobsoni* (Oud.) against tau-fluvalinate. *Exp. & Appl. Acarol.,* 20 (5): 283-296

Ho, K.-K. and J.K. An. 1980. [Effects of Gubitol and its application methods on honeybee mite (*arroa jacobsoni* Oudemans) in Taiwan]. *Honeybee Science* 1: 155-156. In Japanese, English summary.

Houck, A.M., J.B. Clark, K.R. Peterson, M.G. Kidwell. 1991. Possible horizontal transfer of *Drososphila* genes by the mite *Proctolaelaps regalis. Science.* 253: 1125-1229.

Imdorf, A., V. Kilchenmann, S. Bogdanov, B. Bachofen and C. Beretta. 1995. Toxizität von Thymol, Campher, Menthol und Eucalyptol auf *Varroa jacobsoni* Oud und *Apis mellifera* im Labortest. *Apidologie* 26: 27-31.

Jelinski, M., A. Jedruszuk and R. Kostecki. 1994. Studies on the usefulness of different varroa control substances incorporated in plastic carriers. In: New perspectives on varroa, Proc. Intern. Meet., Rez near Prague 1993, A. Matheson ed., IBRA, Cardiff, 85-86.

Kilani, M., J. Bussiéras, A. Popa and A. Sakli. 1981. Essai préliminaire de traitement de la varroase (à *Varroa jacobsoni*) de l'abeille domestique par l'amitraz. *Apidologie* 12: 31-36.

Koeniger, N. 1986. Transport und Verteilung von Akariziden im Bienenvolk unter Ausnutzung des natürlichen Körperkontaktes der Bienen. *Apidologie* 17: 381-383.

Koeniger, N., A. Klepsch and V. Maul. 1983. Zwischenbericht über den Einsatz von Milchsäure zur Bekämpfung der Varroatose. *Allg. dtsch. Imkerztg.* 17: 209-211.

Künzler, K., H. Mook, and H. Breslauer. 1979. Untersuchung über die Wirksamkeit der Ameisensäure bei der Bekämpfung der Bienenmilbe *Varroa jacobsoni. Biene* 115: 372-373.

Lodesani, M., A. Pellacani, S. Bergomi, E. Carpana, T. Rabitti T. and P. Lasagni. 1992. Residue determination for some products used against *Varroa* infestation in bees. *Apidologie* 23: 257-272.

Loglio, G. and G. Plebani. 1992. Valutazione dell'efficacia dell'Apistan. *Apic. mod.* 83: 95-98.

Matheson, A. 1995. First documented findings of *Varroa jacobsoni* outside its presumed natural range. *Apiacta* 30: 1-8.

Mikityuk, V.V. and O.F. Grobov. 1978. [Thymol for varroatosis]. *Pchelovodstvo* no. 3, 26.

Milani, N. 1993. Analytical bibliography on *Varroa jacobsoni* Oud. and related species. *Apicoltura* 8: Ap-

pendix, 147 pp.

Milani, N. 1994. Possible presence of fluvalinate-resistant strains of *Varroa jacobsoni* in northern Italy. *In:* New perspectives on varroa, Proc. Intern. Meet., near Rez near Prague 1993, A. Matheson ed., IBRA, Cardiff, 87.

Milani, N. 1995. The resistance of *Varroa jacobsoni* Oud. to pyrethroids: a laboratory assay. *Apidologie* 26: 415-429.

Milani, N. 1996. Die Ausbreitung apistanresistenter Varroamilben im nordöstlichen Italien. *Bienenvater* 117: 290-193.

Milani, N. and G. Della Vedova. 1996. Determination of the LC_{50} in the mite *Varroa jacobsoni* of the active substances in Perizin® and Cekafix®. *Apidologie* 26: 67-72.

Popov, E.T. 1983. [Oxalic acid for varroatosis control]. *Pchelovodstvo* No. 7, 13. In Russian.

Rickli, M., A. Imdorf and V. Kilchenmann. 1991. Varroa-Bekämpfung mit Komponenten von ätherischen Ölen. *Apidologie* 22: 417-421.

Ritter, W. 1985. First results from biological trials with Apitol: a medicament with systemic activity. Proc. XXXth Intern. Congr. Apic., Nagoya 1985, Apimondia Publ. House, Bucharest, 189-190.

Ritter, W. and H. Roth. 1988. Experiments with mite resistance to varroacidal substances in the laboratory. European research on varroatosis control, Proc. Meet. EC Experts' Group, Bad Homburg 1986, R. Cavalloro ed., Balkema, Rotterdam, 157-160.

Ronzoni, C. 1988. Residui di Amitraz nel miele? *Ape nostra amica* 10: No. 2, 29-31.

Schäbitz, H., N. Koeniger and S. Fuchs. 1991. Wirksamkeit und Bienentoleranz von Cekafix® sowie ein Vergleich mit Perizin® in Labortests. *Apidologie* 22: 468-470.

Schieferstein, E. 1985. Nachdrückliche Warnung vor der Anwendung von Amitraz! *Neue Bienenzucht* 11: 378.

Senegacnik, J. 1991. O zatiranju varoze z vodno emulzijo fluvalinata [On the control of varroatose with water emulsion of fluvalinate]. *Zb. Vet. Fak. Univ. Ljubljana* 28: 25-32. In Slovene.

Stoya, W., G. Wachendörfer, I. Kary, P. Siebentritt and E. Kaiser. 1986. Ameisensäure als Therapeutikum gegen Varroatose und ihre Auswirkungen auf den Honig. *Dtsch. Lebensm.-Rundsch.* 82: 217-221.

Taccheo Barbina, M. and M. De Paoli. 1994. Degradation in the laboratory, and residues in honey and wax samples from field trials, of flumethrin. In: New perspectives on varroa, Proc. Intern. Meet., Rez near Prague 1993, A. Matheson ed., IBRA, Cardiff, 91-96.

Taccheo Barbina, M., M. De Paoli, F. Chiesa, M. D'Agaro and U. Pecol. 1989. Coumaphos decay and residues in honey samples. *In:* Present status of varroatosis in Europe and progress in the varroa mite control, Proc. Meet. EC-Experts' Group, Udine 1988, R. Cavalloro ed., C.E.C., Luxembourg, 379-386.

Taccheo Barbina, M., M. De Paoli, S. Marchetti and M. D'Agaro. 1988. Bromopropylate decay and residues in honey samples. *In:* European research on varroatosis control, Proc. Meet. EC Experts' Group, Bad Homburg 1986, R. Cavalloro ed., Balkema, Rotterdam, 131-143.

van Buren, N.W.M., A.G.H. Mariën and H.H.W. Velthuis. 1992. The role of trophallaxis in the distribution of Perizin in a honeybee colony with regard to the control of *Varroa* mite. *Entomol. exp. appl.* 65: 157-164.

Vesely, V. 1993. Acrinathrin gegen *Varroa jacobsoni*. *Apidologie* 24: 499-500.

Vesely, V., M. Machova, J. Hessler, V. Hostomská and J. Leniocek. 1994. Reduction of fluvalinate residues in beeswax by chemical means. *J. Apic. Res.* 33: 185-187.

Wachendörfer, G., E. Kaiser, K. Krämer and D. Seinsche. 1983. Labor- und Feldversuche mit einem von Krämer modifizierten Ameisensäure-Dämmplatten-Verfahren zur Varroatosebekämpfung. *Allg. dtsch. Imkerztg.* 17: 339-344.

Wallner, K. 1995. Nebeneffekte bei Bekämpfung der Varroamilbe. Die Rückständssituation in einigen Bienenprodukte. *Bienenvater* 116: 172-177.

Wallner K. 1996. Rückstände in Bienenprodukten. Die zentrale Rolle des Bienenwachses. *Allg. dtsch. Imkerztg.*, 30, no. 6, 10-14.

Wallner, K., M. Luh, R. Womastek, H. Pechhacker and R. Moosbeckhofer. 1995. Entwicklung der Rückstandssituation am Institut für Bienenkunde seit beginn der Varroabehandlung. *Bienenvater* 320-329.

Watkins, M. 1996. Resistance and its relevance to beekeeping. *Bee World* 77 (4): 15-22.

Wienands, A. 1988. Synopsis of the worldwide applied preparations against *Varroa* disease of honeybees. A. Wienands, Bonn, 6th ed., 7 pp.

Keywords: *Varroa jacobsoni,* honey bee, treatment, residues, resistance, fluvalinate.

In the shadow of varroa—tracheal mites

DIANA SAMMATARO
DEPT. ENTOMOLOGY, OARDC BEE LAB
1680 MADISON AVE.
WOOSTER OH 44691

Tracheal mites or acarine disease

Acarine disease, now known by its causative agent the tracheal mite, was discovered in the Isle of Wight in the 1920's and has since spread to all areas where European honey bees are kept, with the exception of Australia and Oceania. The small mite was later named (*Acarapis woodi* [Rennie]) and lives inside the tracheae of adult bees. A newly-mated honey bee tracheal mite (HBTM) female emerges from an old host bee and, by crawling up on the bee's hair or seta, quests to find a newly emerged or callow bee; see Figure 1. Once another bee comes close, the mite will transfer to the new host.

HBTM can discriminate between old and young hosts by the cuticular hydrocarbons on the bee's cuticle (Phelan *et al.* 1991). Once the mite finds a suitable host, she enters the trachea by means of the spiracle opening, and can lay 0.85 eggs/day

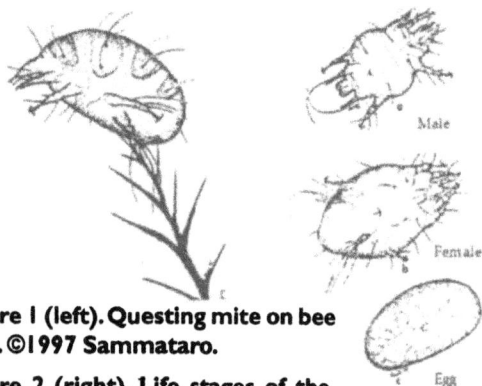

Figure 1 (left). Questing mite on bee seta.©1997 Sammataro.

Figure 2 (right). Life stages of the tracheal mite.©1997 Sammataro

Sammataro: Honey bee tracheal mites

for eight to 12 days. After the eggs hatch, the immature mites, or *larvae*, live as parasites until eclosion. All mite stages, except the egg, live inside all castes of adult bees, feeding on bee hemolymph by piercing the walls of the tracheal tubes with their style-like chelicerae; see Figure 2.

New mites emerge 11-12 days (males) or 14-15 days later (females) after the egg was laid. HBTM can cause severe bee losses, especially where bees are confined for several months in cold climates.

Distribution of HBTM

Distribution of HBTM is now worldwide. First reported in the United States in honey bees from Texas apiaries in 1984, this mite is responsible for significant colony losses throughout North America (Delfinado-Baker 1988). Prior to mite detection, 11% of colonies perished over the winter in most northern states, but in Pennsylvania, those losses increased to 31% due to tracheal mite infestation (Sammataro 1995; Tomasko *et al.* 1993). In general, mite populations increase when bees are confined to their hive during the winter, and decrease in the summer when bee populations are highest (Dawicke *et al.* 1992). As yet, the relationship between mite-infested colonies and colony death is unclear. From its first report, the mite has now been spread by migratory beekeepers and queens or packages into many regions of the beekeeping world.

The evolutionary history of tracheal mites is still unknown. Only a few mites inhabit the tracheal systems of arthropods (Sammataro 1995), and Eickwort (1993) speculated that *Acarapis* evolved from saprophagous or predatory mites. One reason for their appearance in bee hives may be the nesting behavior of Apinae: two species (*Apis mellifera* and *A. cerana*) evolved to nest in cavities, which at the same time provided habitats for such mites. While the whole-colony effects of tracheal mites have been studied by many, not yet understood is how minute, eyeless mites leave their original hosts, negotiate the terrain of their old hosts and locate an oviposition site in a new bee's trachea.

Some countries report that tracheal mites are no longer a problem to beekeepers, citing that all the susceptible bees have died off (see HBTM Worldwide Reports, below), or the lethal strains of mite have modified. However, where winter is severe, colony losses may still be high. With the introduction of varroa, many beekeepers no longer investigate the incidence of HBTM. However, their effect may still be seen in unexplained colony loss where varroa is controlled.

Recent work done by bee researchers in Pennsylvania indicate that many queen breeders from the southern states or California, where HBTM are not considered a problem, are no longer treating for HBTM. As a result, queens are being delivered infested with the mite, which appear to be resulting in poor queen performance, high supersedure rates and queen 'disappearance'. Some queens are already 'heavily' infested when ten to 14 days old.

Testing for tracheal mites

External signs are extremely unreliable, but include dwindling populations of bees, weak bees crawling on ground with K-wings (see Chronic Bee Paralysis virus, below), and abandoned hives in the spring with plenty of honey stores. In order to determine if you have mites, bees MUST be dissected; the collection procedure is outlined below:

• Sample at least 50 percent of the colonies in the apiary.

• Collect only "old" bees; they are most likely to have an infestation and are the easiest to diagnose (see below).

• Old bees are usually on the inner cover, at the entrance or out foraging, *not* near the broodnest.

• Collect at least 50 bees per colony and place them in a 70 percent ethanol or isopropyl solution, or freeze them in a glass or plastic jar or bag.

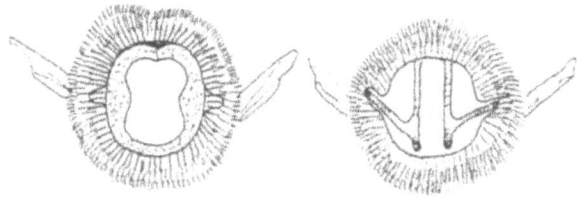

Figure 3. Left, ventral view: bee with head removed, showing collar. Right, bee ventral view, with collar removed, showing prothoracic tracheal tubes. ©1997 Sammataro.

A positive diagnosis of the tracheal mite by gross examination of the colony or by bees walking around on the ground *cannot* be done. Some of the visible symptoms are not always reliable and are not necessarily due to HBTM. A dissecting microscope (at about 40 to 60x) and a pair of fine jeweler's forceps are needed to do the examination. The dissection procedure is outlined below:

Soften a frozen bee by holding it in your hand a few seconds.

If the bee was stored in ethanol, it is soft enough but the tissues will be darkened, even after one month.

Place the bee on its back and pin it on a piece of corkboard or a wax-impregnated petri dish, through the thorax, between the second and third pairs of legs. You can also hold the bee in your fingers and proceed; do this under the microscope.

Remove the head and pull up the collar surrounding the thoracic opening with the forceps; see illustration.

The thoracic trachea will be exposed when this covering is removed—in a healthy bee it looks pearly white with inner coils, like a dryer hose.

If mites are present, the trachea will have shadows or be spotted—the spots are all stages of mites. In severe infestations, the tube can be completely brown or black.

Darkened tracheae will be visible to the naked eye, while healthy tracheae will be white and shiny. You can use this method to detect heavy infestations (spring and fall) but not light ones, such as in the summer.

HBTM & virus

There is only one virus associated with tracheal mites and that is the **Chronic Paralysis virus** (CPV) first reported in 1933. It may have been the "Isle of Wight" disease, as the symptoms are very similar to bees dying of heavy HBTM infestation. CPV comes in two syndromes: *Type I* is distinguished by bees trembling, unable to fly, with K-wings and distended abdomens. *Type II*, called the hairless-black syndrome, is recognized by hairless, black shiny bees crawling in front of the hive.

A wound as small as a broken bee hair is enough to transfer this disease, as well as having contaminated hairs mixed in with bee food, such as pollen patties. It also is prevalent when bees are confined to the colony for a long period of time. To date, no clear association has been found between bees, mites and CPV. However, virus-like picarnovirus particles were found in tracheal mites from the UK (Liu 1991).

Controlling tracheal mites

In the USA, menthol crystals, from the plant *Mentha arvensis*, is sold in crystal form (98%) at many bee supply stores and each two story colony takes 1.8 oz. (50 g) of the menthol or one packet. The problem with menthol is that it is temperature dependent, and the odor will sometimes make bees leave the hive if the temperature is too hot, or be ineffective if the temperature is too cold. It should be in the colony at least two weeks. Remove all menthol at least one month before the surplus honeyflow to keep honey from becoming contaminated.

An alternate method is to use vegetable oil or vegetable shortening patties; a shortening and sugar patty kept on all the time seems to protect against these mites (Sammataro *et al.* 1994). Patties are made by mixing two parts white granulated sugar to one part solid vegetable shortening (do NOT use animal fat). Place one, quarter pound oil/sugar patty, about the size of your hand, on the top bars at the center of the broodnest. The bees at most risk are the young nurse bees, found in the broodnest. The patty should last about a month, then replace it with another oil patty. If some colonies appear to remove the patty much faster, they may be displaying hygienic behavior; try rearing queens from them. Since young bees are continually emerging, it is important to have the patty present in the colony for an extended time. The best time to treat is when mite levels are climbing, i.e. in the fall and early spring.

Treatment and prevention schedule

To help reduce infestations levels of these mites, researchers have found the following steps to give good control.

a. Late Summer, after honey supers are removed: Requeen the hive and reduce field force. Old bees and drones have high infestations of tracheal mites. By requeening and reducing the number of old bees, a new queen can go into the winter with bees that have a reduced mite load and will be able to make it to spring in better condition.

Place capped/uncapped brood and house bees plus a new queen into a new hive and either move them to a new location or place the box behind the old hive. Leave older bees and drones behind or eliminate with a soapy water spray. (If you don't want to do this, fumigate the older colony with menthol crystals for the prescribed time, and join back to the new queen later). Place an oil or extender patty on the top bars, above the broodnest, of the new colony; make sure many bees come in contact with it. Treat all hives with Apistan strips and patties or menthol which will reduce the infestation rates of both mites to a non-fatal level.

b. Early Fall: In 30 to 45 days, remove the strips and menthol packet. Check the colonies for food stores and feed fumigillan to control Nosema. If the new queen is laying well, join the split back to the old hive that has been treated with menthol and Apistan. Keep or replace the patty as needed. This is a crucial time for tracheal mites when bees are clustered together; the mites transfer easily to bees when they are close together. The patty insures protection for young bees and reduces mite levels.

c. Late fall: Check bees for food stores and feed syrup or clean honey if needed. Keep or replace patty.

d. Winter: Follow recommended practices for wintering colonies in your area. These may include wind breaks, hive ventilation, wrapping or other preparations. Make sure food stores are ample and patty is present.

e. Spring: Check food stores and patty; replace or feed as needed. When weather starts to break, feed pollen supplements or substitutes. As bees forage, keep alert for the first signs of varroa mites and treat accordingly.

These steps should help reduce mite levels. Experiment with techniques that are common in your area. Talk to other beekeepers, go to bee meetings, compare notes with others and read some good bee journals to keep up with the new research and treatments for mites.

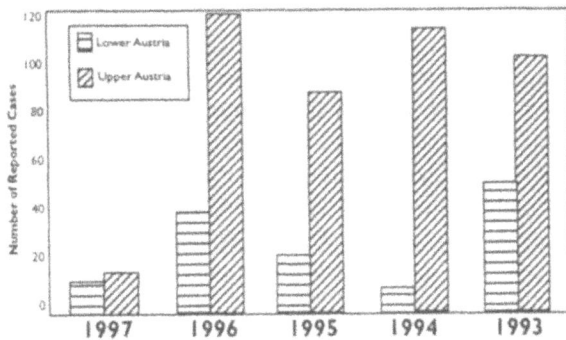

Figure 4. Tracheal mites in Austria. Data supplied by I. Derakhshifar.

HBTM—worldwide reports

Austria (see Figure 4)

A beekeeper with about 200 colonies detected a high HBTM infestation in the spring of 1995 by visible signs (crawling bees, K wing etc.). The investigation revealed 60% of the bees were infested. He treated with formic acid which was very effective. HBTM are found more in colder areas of Austria.

Rudolf Mossbeckhofer
Institut fuer bienenkunde,
A-1226, POB 400, Spargelfeldstrasse 191.
1220 Wien, Austria
Email: *moosbeck@bfl-sg5.bfl.gv.at*

Finland (see Figure 5)

Since 1991 infestation have been found every year for HBTM, totaling 29% of beekeepers in Finland. Infestations started from 1991 imports of queens

Figure 5. Tracheal mites in Finland 1997. Data supplied by S. Korpela

Sammataro: Honey bee tracheal mites

from USA. A large loss of bee colonies occurred in the winter or 1995-6 in the apiaries of the largest queen breeder in Finland (he lost 20% of his bees).

Losses of bee colonies have been variable, depending on the stage when the infestation was detected and the percentage of the infested colonies in the apiaries. Some losses are widespread, some not. For control I have recommended formic acid and the results show that infestations can be controlled.

I have received articles from Dr. Ritter in Germany where he writes that the HBTM is widespread in SW parts of Germany, but causes almost no losses. He explains that the mite can be controlled naturally by supplying the bees good foraging conditions in spring so that bees can 'outgrow' the mite populations.

The reason for mite disappearance seems to be in warm springs when bees could fly and rear brood. On the other hand, last spring and the first half of summer was extremely cold and wet so this triggered a rapid increase of mite populations instead of disappearing. We lost, this past winter, 10 out of 14 colonies due to HBTM.

Seppo Korpela
Ag. Research Center of Finland
Institute of Plant Protection,
FIN-31600 Jokioinen, Finland
Phone: 358 3 4188 576, Fax: 358 3 4188 584
Email: *seppo.korpela@mtt.fi*

France

About your question on current status of tracheal mites in Europe, I do not experience this mite, but I just wanted to tell you that this mite is not a problem anymore in France for a few years. We do not treat the colonies for that, and actually it could be very difficult to find this mite if we would need it for experiments.

Yves Le Conte
I.N.R.A., Unité de Zoologie, Laboratoire de Biologie de L'Abeille
Domaine St Paul, Site Agroparc,
84914 Avignon Cedex 9, France
Tel: (33) 04.90.31.62.18
Fax: (33) 04.90.31.62.70
Email: *leconte@avignon.inra.fr*

Germany

In the years around 1950 to 1960 German bee researchers and veterinarians killed about 10.000 colonies because of tracheal mite. Those researchers claimed to have eradicated acarapis in some regions. We checked this and found acarapis in Kronberg in 1968, which was reported to be mite free 10 years ago. We found mites in dead winter bees. Because of the legislation (all colonies had to be reported to the veterinary service and were killed) we did not put it down in protocol. The mite legislation is the main reason for not having proper scientific publications. Everybody had realized that these mites were harmless here, very much in contrast to the fierce actions of the law. So everybody stopped looking; only very recently, acarapis regulations were changed.

Email: *Nikolaus.Koeniger@em.uni-frankfurt.de* (nikolaus koeniger)

Holland

Since 1945 until 1984 in Holland the stamping out method was used in case *Acarapis woodi* was found in honeybees. In the early eighties we transported about 50 infected colonies to a North Sea island (Ameland), in order to test the effect of tobacco smoke (used for Varroa diagnosis at the time) and Folbex VA (the first drug available against Varroa) on Acarapis mites. Colonies that had over 30% infection died during the first winter. The rest of the colonies (treated and untreated!) recovered and after two years we could not find a single mite. If the circumstances are good, Acarine disease can disappear without any treatment. This is why we stopped searching for mites and abandoned the stamping out method. In the following years no infections were found (in samples of dead bees sent in by beekeepers). Maybe this has something to do with the widespread use of the systemic acaricide Perizin against Varroa. In recent years Apistan strips have replaced Perizin treatment. This year for the first time again we found three infected apiaries after inspection of samples of dead bees sent in by beekeepers, in different parts of the country. The colonies had been treated with Apistan last autumn. *Acarapis* is still present in low numbers and can become a problem after a bad summer. You can download some pictures of *Acarapis* and *Varroa* from our homepage.

Aad de Ruijter
Research Centre for Insect Pollination and Beekeeping "Ambrosiushoeve"
E-mail: *ambros@pi.net*
homepage:
http://home.pi.net/~ambros/home.html

Italy

In Italy tracheal mites are seldom reported as a cause of damage. The presence of infested bee colonies should be reported to the veterinary services, but it is unlikely that official statistics (if there are such data!) give a realistic picture of the situation. We never carried out investigations on the acariosis. Just to give you an idea, I can give you some data from our "Laboratorio apistico regionale". In our small region (20,000 bee colonies, some 2,000 apiaries, about 2% of the total bee colonies of Italy) there are some 30 qualified, experienced beekeepers, who report suspect cases and send samples to our Department. In 1995, we analyzed 7 samples; only one revealed the presence of a weak infestation. In 1996, again we received 7 samples, only one was positive (the mite was detected in only one colony out of three from the same apiary). Only in 1994 we detected a strong infestation in one apiary, in nuclei bought elsewhere and showing clear signs of damage. Five further samples from other apiaries were negative.

Norberto Milani
Dipartimento di Biologia applicata alla Difesa delle Piante
Università degli Studi di Udine
Via delle Scienze, 208, I 33100 UDINE (Italy)
Tel +39 (0)432 55 85 08 - 55 85 03 (secr.)
Fax +39 (0)432 55 85 01
Email: *Norberto.Milani@Pldef.Uniud.it*

South Africa

In 1995, HBTM was reported in S. Africa (see Buys 1995) and is reported to be of minor economic importance at present. Thank you very much for the information. Yes, I am still looking for tracheal mites, as a matter of fact, Ben Buys sent me potentially infested bees and I am busy looking for mites.

Eddie Ueckermann
Acarologist
ARC-Plant Protection Research Institute
South Africa
Email: *rieteau@PLANT2.AGRIC.ZA*

Sweden

On your recent request for information on tracheal mites, we recently did a survey nationwide in Sweden for this parasite. It has never been detected before, and we did not find it now (slicing tens of thousands of bees). It was once detected in Norway and from 100 queens in Denmark.

Ingemar Fries, Department of Entomology
Bee Division, Swedish Univ. of Agric. Sci.
Box 7044, S-750 07 Uppsala, Sweden
Fax: Int+ 46 18 67 28 90
Tel: Int+ 46 18 67 20 73
E-mail: *Ingemar.Fries@entom.slu.se*

Switzerland

Research since the 1980's on the effect of HBTM were carried out. While infestation is high in winter, by spring mites disappear. Mites do not seem to destroy bee colonies but losses were due to honeydew honey with it's high mineral content. See report by Wille 1987 in References.

Peter Fluir
Eidgenössische Forschungsanstalt für Milchwirtschaft
Sektion Bienen
CH- 3097 Liebefeld, Bern

United Kingdom

Tracheal mite (called acarine disease in the UK for some strange reason, or acariosis/acariasis elsewhere in Europe) does not seem to be a big issue at the moment. It is sporadic throughout Europe at most and although it is notifiable in a few European states it isn't regarded as serious overall and so there aren't many good records of incidence. There are a few countries with quite a low incidence as far as I can gather, *e.g.*, Denmark and Holland. Tracheal mite is fairly common, though, especially in the south of England (more beekeepers) and does cause occasional local difficulty for some beekeepers. There is some evidence that recent (*i.e.*, in the last few years) imports of 'Italian-type' bees from New Zealand (we import a few thousand each year) are prone to the mite. From our own experience this would seem to be true. We used them ourselves because they were good to handle and very productive in good years. But winter losses were very high and nearly always associated with high tracheal mite counts. The problem disappeared (although the mite remains at low levels) when we reverted to a mixture of local 'mongrels' and Buckfast strains.

Medwin H. Bew, Head of Unit
Central Science Laboratory
National Bee Unit, Room 10GA03
Sand Hutton, YORK
England YO4 1LZ
Tel: (0)1904 462506
Fax: (0)1904 462240
Email: *m.bew@csl.gov.uk*

References for tracheal mites

1984. *Acarapis woodi* in the United States. *Am. Bee J.* 124: 805-806.

Acarapis woodi in the United States. *Am. Bee J.* 124: 805-806. 1984.

Adam, Brother (1968) "Isle of Wight" or acarine disease: its historical and practical aspects. *Bee Wld.* 49 (1), 6-18.

Allen E. 1989. Looking for tracheal mites – how, when and why. *Am. Bee J.* 129 (6), 396-398.

Allen, M.R. & B. V. Ball. 1996. The incidence and world distribution of honey bee viruses. *Bee World* 77: 141-162.

Bailey, L. 1956. The effect of *Acarapis woodi* on honey bees from North America. *J. Apicult. Res.* 4, 105-108.

Bailey, L. 1958. The epidemiology of the infestation of the honey bee, *Apis mellifera* L. by the mite *Acarapis woodi* Rennie and the mortality of infested bees. *Parasitology* 48, 493-506.

Bailey, L. 1961. The natural incidence of *Acarapis woodi* (Rennie) and the winter mortality of honey bee colonies. *Bee Wld.* 42, 96-100.

Bailey, L. 1963. Infectious Diseases of the Honey-Bee. London: Land Books.

Bailey, L. 1964. The "Isle of Wight Disease": the origin and significance of the myth. *Bee Wld.* 45:32-37.

Bailey, L. 1965. The effects of *Acarapis woodi* on honeybees from North America *J. Apic. Res.* 4, 105-108.

Bailey, L. 1967. The incidence of *Acarapis woodi* in North American strains of honey bee in Britain. *J. Apic. Res.* 6, 99-103.

Bailey, L. 1985. *Acarapis woodi* : a modern appraisal. *Bee Wld.* 66: 99-104.

Bailey, L. & D. C. Lee. 1959. The effect of infestation with *Acarapis woodi* (Rennie) on winter mortality of honey bees. *J. Insect Pathol.* 1, 15-24.

Bailey, L. & J. N. Perry. 1982. The diminished incidence of *Acarapis woodi* (Rennie) (Acari: Tarsonemidae) in honey bees, *Apis mellifera* L. (Hymenoptera: Apidae) in Britain. *Bull. Entomol. Res.* 72: 655-662.

Bailey, L. 1989. Some notes on *Acarapis woodi* (Rennie). *Am. Bee J.* 129 (8), 543-545.

Bailey, L., B.V. Ball, J. M. Carpenter & R. D. Woods. 1980. Small virus-like particles in honey bees associated with chronic paralysis virus and with a previously undescribed disease. *J. Gen. Virol.* 46: 149-155.

Ball, B. 1994. Host-parasite-pathogen interactions. In Matheson, A. Ed. *New perspectives on varroa.* IBRA, Cardiff, Wales, 5-11.

Bioenvironmental Bee Laboratory. 1984. Diagnosis of acarine disease. Leaflet. Beltsville, MD.

Bruce, W.A., K.H. Hackett, H. Shimanuki & R.B. Henegar. 1991. Bee mites: vectors of honey bee pathogens? In Prod. Apimondia Symp. on Recent Res. on Honey Bee Path. Gent, Belgium, Sept 5-7, 1990 (Ritter, W. ed.) Janssen Pharmaceutica, Beerse, Belgium, 180-182.

Buys, B. 1995. First record in South Africa of the tracheal mite *Acarapis woodi. S. Afr. Bee J.* 67: 75-80.

Calderone, N.W. & H. Shimanuki. 1992. Evaluation of sampling methods for determining infestation rates of the tracheal mite (*Acarapis woodi* R.) in colonies of the honey bee (*Apis mellifera*): spatial, temporal and spatio-temporal effects. Exp. & Appl. Acarol. 15: 285-298.

Calderone, N. W. & H. Shimanuki. 1995. Evaluation of four seed-derived oils as controls for *Acarapis woodi* (Acari: Tarsonemidae) in colonies of *Apis mellifera* (Hymenoptera: Apidae). J. Econ. Entomol. 88(4): 805-809.

Camazine, S. 1985. Tracheal flotation: a rapid method for the detection of honey bee acarine disease. Am. Bee J. 125 (2), 104-105.

Clark, K.J. 1991. Winter survival times of colonies infested at various levels by tracheal mite. Proc. Amer. Bee Res. Conf., Am. Bee J. 131 (12),773.

Clark, K.J. & J. Gates. 1991. Tracheal mite control trials in British Columbia. Proc. Amer. Bee Res. Conf., Am. Bee J. 131 (12), 773.

Colin, M.E., Faucon, J.P, Giauffret, A. & C. Sarrazin. 1979. A new technique for diagnosis of acarine infestation in honeybees. J. Apicult. Res. 18 (3), 222-224.

Conner, L. Tracheal mites: history, biology and control. 1987. Glean. Bee Cult. 115 (9), 516-519,520.

Cox, R.L., J.O. Moffett, W.T. Wilson and M. Ellis. 1989. Effects of late spring and summer menthol treatment on colony strength, honey production and tracheal mite infestation levels. Am. Bee J. 129 (8), 547-549.

Cox, R.L., J.O. Moffett, W.T. Wilson. 1989. Techniques for increasing the evaporation rate of menthol when treating honey bee colonies for control of tracheal mites. Am. Bee J. 129 (2) 129-131.

Cox, R.L., W.T. Wilson, & D. L. Maki. 1986. Chemical control of the honey bee tracheal mite, *Acarapis woodi*. Am. Bee J. 126, 828.

Cromroy, H. L., J.C. Nickerson, D.L. Harris & L.P. Catts. 1986. Research and control studies on the tracheal mite in Florida. Proc. Honey Bee Tracheal Mite (*Acarapis woodi* R.) Sci. Symp., Am. Assoc. Prof. Apic. and Can. Assoc. Prof. Apic., 48.

Danka, R. G., J. D. Villa, *et al.* Field test of resistance to *Acarapis woodi* (Acari: Tarsonemidae) and of colony production by four stocks of honey bees (Hymenoptera: Apidae). J. Econ. Ent. 88: 584-591. 1995.

Dawicke, B. L., G. W. Otis, C. Scott-Dupree & M. Nasr. 1992. Host preference of the honey bee tracheal mite (*Acarapis woodi* (Rennie)). Exper. & Appl. Acarol. 15: 83-98.

De Jong D, R.A. Morse & G.C. Eickwort. 1982. Mite pests of honey bees. Ann. Rev. Entomol. 27, 229-252.

Delaplane, K. S. 1992a. Controlling tracheal mites (Acari: Tarsonemidae) in colonies of honey bees (Hymenoptera: Apidae) with vegetable oil and menthol. J. Econ. Ent. 85 (5): 2118-2124.

Delaplane, K. S. 1992b. Controlling Tracheal Mite. Am. Bee J. 132: 577-578, 611.

Delfinado-Baker, M. 1984. *Acarapis woodi* in the United States. Am. Bee J. 124 (11), 805-806.

Delfinado-Baker, M. & E.W. Baker. 1982. Notes on honey bee mites of the genus *Acarapis* Hirst (Acari: Tarsonemidae). Internat. J. Acarol. 8 (4), 211-226.

Delfinado-Baker, M. 1988. The tracheal mite of honey bees: a crisis in beekeeping. In Needham, G.R., R.E. Page, Jr., M. Delfinado-Baker, & C.E. Bowman eds.

African Honey Bees and Bee Mites. Ellis Horwood, Ltd. Chichester, 493-497.

Eastern U.S. Tracheal Mite Symposium. Abstracts from Nov. 30 Dec.1, 1989 meeting of Empire State Honey Producers Assoc. Syracuse, NY. Am. Bee J. 130 (3) 185-187.

Eckert, J. E. Acarapis mites of the honey bee, (*Apis mellifera* Linnaeus). J. Insect Pathology, 3: 409-425. 1961.

Eickwort, G.C. 1990. Associations of mites with social insects. Ann. Rev. Entomol. 35, 469-488.

Eickwort, G.C. 1993. Evolution and life-history patterns of mites associated with bees. In: Mites . Ecological and Evolutionary Analyses of Life-History Patterns. (Ed. by M. A. Houck), pp. 218-251. Ithaca New York: Chapman & Hall.

Eischen F.A., D. Cardoso-Tamez, A. Dietz & G.O. Ware. 1988. Cymiazole, a systemic acaricide that controls *Acarapis woodi* (Rennie) infesting honey bees. I. Laboratory Tests. II. Apiary Tests. Apidologie, 19 (4), 367-376.

Eischen F.A., W.T. Wilson, D. Hurley & D. Cardoso-Tamez. 1988. Cultural practices that reduce populations of *Acarapis woodi* (Rennie). Am. Bee J. 128 (3), 209-211.

Eischen, F. A. & A. Dietz. 1986. *Acarapis woodi* studies in northeastern Mexico. Proc. Honey Bee Tracheal Mite (*Acarapis woodi* Rennie) Sci. Symp., Am. Assoc. Prof Apic. & Can. Assoc. Prof. Apic., 52-53.

Eischen, F. A. 1987. Overwintering performance of honey bee colonies heavily infested with *Acarapis woodi* (Rennie). Apidology 18 (4): 293-304.

Eischen, F.A., D. Cardoso-Tamez, W.T. Wilson, & A. Dietz. 1989. Honey production of honey bee colonies infested with *Acarapis woodi* (Rennie). I. Apidology 20, 1-8.

Eischen, F.A., J.S. Pettis & A. Dietz. 1987. A rapid method of evaluating compounds for the control of *Acarapis woodi* (Rennie). Am. Bee J. 127 (2), 99-102.

Ellis, M.D. & C. Simonds. 1987. Menthol fumigation of caged honey bees to prevent the spread of mites in package and queen bee shipments. Am. Bee J. 127, 844.

Francis, B.R., W.E. Blanton, J.L. Littlefield & R.A. Nunamaker. 1989. Hydrocarbons of the cuticle and hemolymph of the adult honey bee (Hymenoptera: Apidae). Ann. Ent. Soc. Am. 82 (3), 486-494.

Furgala B., S. Duff, S. Aboulfaraj, D. W. Ragsdale & R.A. Hyser. 1988. Some effects of the honey bee tracheal mite (*Acarapis woodi* Rennie) on non-migratory, wintering honey bee (*Apis mellifera* L.) colonies in east central Minnesota. Am. Bee J. 129 (3),195-197.

Garcia, D. 1991: Ameisensure gegen Tracheenmilbe. In Freilandversuchen in Nordost-Mexiko als wirksam erwiesen. Deutsches Imker-Journal, Heft 11, 1991, 455-456.

Gary, N. E. & R. E. Page Jr. 1987. Phenotypic variation in susceptibility of honey bees, *Apis mellifera*, to infestation by tracheal mites, *Acarapis woodi*. Exp. & Appl. Acarol. 3: 291-305.

Gary, N. E., R. E. Page Jr. 1989. Tracheal mite (Acari: Tarsonemidae) infestation effects on foraging and survivorship of honey bees (Hymenoptera: Apidae). J. Econ. Entomol., 82 (3), 734-739.

Gary, N. E., R. E. Page Jr. & K. Lorenzen. 1989. Effect of age of worker honey bees (*Apis mellifera*) on tracheal mite (*Acarapis woodi*) infestation. *Exp. & Appl. Acarol.* 7 :153-160.

Gary, N.E., & R.E. Page, Jr. 1988. Factors that affect the infestation of worker honey bees by tracheal mites, (*Acarapis woodi*). 506-511. in Needham, G.R., R.E. Page, Jr., M. Delfinado-Baker, & C.E. Bowman eds. *African Honey Bees and Bee Mites.* Ellis Horwood, Ltd. Chichester.

Gary, N.E., R. E. Page Jr., R.A. Morse, C. E. Henderson, M. E. Nasr & K. Lorenzen. 1990. Comparative resistance of honey bees (*Apis mellifera* L.) from Great Britain and United States to infestation by tracheal mites (*Acarapis woodi*). *Am. Bee J.* 130 (10), 667-669.

Giordani, G. 1965a. Laboratory research work on *Acarapis woodi* Rennie, the causative agent of the acarine disease of the honeybee *Apis mellifera* L. Note #4. *Bulletin Apicole*, VII No. 2: 159-176.

Giordani, G. 1965b. Laboratory research on *Acarapis woodi* Rennie, the causative agent of acarine disease of honeybees (*Apis mellifera* L.) Note 2. *Bulletin Apicole* 6: 185-203.

Giordani, G. 1965c. Laboratory research work on *Acarapis woodi* Rennie, the causative agent of the acarine disease of *Apis mellifera* L. Note #2. *Vedecké Práce V'yzkumn'ych Ústavú Zemedelsk'ych.* pp. 37-44.

Giordani, G. 1967. Laboratory research work on *Acarapis woodi* Rennie, the causative agent of acarine disease of *Apis mellifera* L. Note #5. *J. Apicult. Res.* 6: 147-157.

Giordani, G. 1977. Course of acarine disease in the field. *Proc. XXVIth. Int. Congr. Apicult.*, Adelaide, Australia: 459-467.

Gruszka, J. 1987. Honey bee tracheal mites: are they harmful? *Am. Bee J.* 127, 653.

Gruszka, J. & D. Peer. 1986. Tracheal mite project at La Ronge, Saskatchewan. *Proc. Honey Bee Tracheal Mite* (Acarapis woodi R.) *Sci. Symp., Am. Assoc. Prof. Apic. and Can. Assoc. Prof. Apic.*, 14-21.

Guzman-Novoa E. & A. Zyzaya-Rubio. 1984. The effects of chemotherapy on the level of infestation and production of honey in colonies of honey bees with acariosis. *Am. Bee J.* 124, 669-672.

Henderson, C. E. & R.A Morse. 1978. Tracheal mites. In: *Honey Bee Pests, Predators and Diseases.* 2nd ed. (Ed. R.A. Morse), ppg. 220-234. Ithaca NY: Cornell Univ. Press.

Herbert, E.W. Jr., H. Shimanuki & J.C. Mathenius, Jr. 1988. An evaluation of menthol placement in hives of honey bees for the control of *Acarapis woodi. Am. Bee J.* 128 (3), 185-187.

Hirschfelder, H. & H. Sachs . 1952. Recent research on the acarine mite. *Bee Wld.* 33, 201-209.

Hirst, S. 1921. On the mites (*Acarapis woodi* (Rennie) associated with Isle of Wight bee disease. *Annals Magazine Nat. Hist.* 7: 509-519.

Hoppe, H., W. Ritter and E.W.- C. Stephen. 1989. The control of parasitic bee mites: *Varroa jacobsoni, Acarapis woodi* and *Tropilaelaps clareae* with formic acid. *Am. Bee J.* 129 (11), 739-742.

Hyser, D. 1986. Tracheal Mite, *Nosema*, and wintered honey bee colonies in Minnesota. *Proc. Honey Bee Tracheal Mite* (Acarapis woodi R.) *Sci. Symp., Am. Assoc. Prof. Apic. and Can. Assoc. Prof. Apic.*, 32-37.

Jadczak, A. M. 1990. Tracheal mites in Maine. *Am. Bee J.* 130: 187.

Jaycox, E.R. Acarine disease of honey bees. 1958. *Cal. Agri. Bull.* 14, 215-221.

Kaliszewski M. & D. L. Wrensch. 1993. Evolution of sex determination and sex ratio within the mite cohort Tarsonemina (Acari: Heterostigmata). In: *Evolution and Diversity of Sex Ratio in Insects and Mites.* (Ed. by D. L. Wrensch & M. A. Ebbert), pp. 192-213. New York: Chapman & Hall.

Killion, E.E. and L.A. Lindenfelser. 1988. Observations on the honey-bee tracheal mites in Illinois, In: *Africanized Honey Bees and Bee Mites* (Ed. by G. R. Needham, R. E. Page, Jr., M. Delfinado-Baker, & C. E. Bowman), pp. 518-525. Chichester: Ellis Horwood, Ltd.

Kjer, D.M., D.W. Ragsdale & B. Furgala. 1989. A retrospective and prospective overview of the honey bee tracheal mite, *Acarapis woodi* R. *Am. Bee J.* 129 (1)Part I: 25-28; Part II: 112-115.

Komeili, A.B. and J.T. Ambrose. 1990. Biology, ecology and damage of tracheal mites on honey bees (Apis mellifera). *Am. Bee J.* 130, 193-199.

Komeili, A.B. and J.T. Ambrose. 1991. Electron microscope studies of the tracheae and flight muscles of noninfested, *Acarapis woodi* infested, and crawling honey bees (*Apis mellifera*). *Am. Bee J.* 131 (4), 253-257.

Lee, D.C. 1963. The susceptibility of honey bees of different ages to infestation by *Acarapis woodi* (Rennie). *J. Insect Pathol.* 5 ? , 11-15.

Lenhert, T., A.S. Michael, & M.D. Levin. 1974. Disease survey of South America africanized bees. *Am. Bee J.* 114 (9), 338.

Lindquist, E.E. 1986. The world genera of Tarsonemidae (Acari: Heterostigmata): a morphological, phylogenetic, and systematic revision, with a reclassification of family-group taxa in the Heterostigmata. In: *Memoirs of the Entomological Society of Canada.* (Ed. A. B. Ewen), pp. 257-262. Ottawa: Entomological Society of Canada.

Liu, T.P. 1990. Ultrastructure of the flight muscle of worker honey bees heavily infested by the tracheal mite *Acarapis woodi. Apidologie.* 21, 537-540.

Liu, T.P. 1991. Virus-like particles in the tracheal mite *Acarapis woodi* (Rennie). *Apidologie.* 22, 213-219.

Liu, T.P., B. Mobus & G. Braybrook. 1989. A scanning electron microscope study on the prothoracic tracheae of the honeybee, *Apis mellifera* L., infested by the mite, *Acarapis woodi* (Rennie). *J. Apic. Res.* 28 (2), 81-84.

Liu, T.P., B. Mobus & G. Braybrook. 1989. Fine structure of hypopharygeal glands from honeybees with and without infestation by tracheal mites, *Acarapis woodi* (Rennie). *J. Apic. Res.* 28 (2), 85-92.

Lozano de Haces, L., W.L. Rubink, W.T. Wilson & M. Guillen-M. 1989. *Nosema* and honey bee tracheal mite interaction in swarms from northeastern Mexico. *Am. Bee J.* 120, 818.

Maki, D.L., W.T. Wilson, and R.L. Cox. 1986. Infestation by *Acarapis woodi* and its effect on honey bee longevity in laboratory cage studies. *Am. Bee J.* 120, -832.

Margolis, L., G. W. Esch, J. C. Holmes, A. M. , & G. A. Schad. 1982. The use of ecological parasitology (Report of an ad hoc committee of the American Society of Parasitologits). *J. Parasitology.* 131-133.

Menapace, D. M. & W .T. Wilson. 1980. *Acarapis woodi* mites found in honey bees from Colombia. *Am. Bee J.*

120 (11), 761-762.

Michale, A.S. 1962. Morphological characters of the honey bee mites. *Bull. Apicole* 4, 21-24.

Ministry of Agriculture, Fisheries and Food. 1956. The examination of bees for acarine disease. Advisory Leaflet Agr. Fisheries, London, No. 362.

Moffett, J. O., R. L. Cox, M. Ellis, R. Rivera, W. T. Wilson, D. Cardoso-T. & J. Vargas-C. 1989. Menthol reduces winter populations of tracheal mites, Acarapis disease. *Bee Wld.* 11, 49-50.

Morgenthaler, O. 1929. Problems of acarine disease of bees. *Bee Wld.* 10, 19-24.

Morgenthaler, O. 1930. New investigations on acarine disease. *Bee Wld.* 11, 49-50.

Morgenthaler, O. 1931. An acarine disease experimental apiary in the Bernese Lake District and some of the results obtained there. *Bee Wld.* 12, 8-10.

Morison, G. D. 1932. A mite (*Acarapis*) that dwells on the back of the honey-bee. *Bee Kingdom,* Leaflet N° 16. 3, 6-11.

Morison, G. D., E. P. Jeffree, L. Murray, & M. D. Allen. 1956. Acarine and *Nosema* diseases of honeybees in Britain, 1925-47. *Bull. Entomol. Res.* 46, 753-759.

Morison, G.D. 1931. Observations on the numbers of mites, *Acarapis woodi* Rennie found in tracheae of the honey bee. *Bee Wld.* 12, 74-76.

Morse, R.A. & G.C. Eickwort. 1990. *Acarapis woodi*, a recently evolved species? Apidmondia, *Proceedings Internt'L Symp. Recent Res. on Bee Path.* W. Ritter, ed. Ghent, Belgium. pp. 102-107.

Mossadegh, M.S. & J.T. Ambrose. 1986. Tracheal mite research in North Carolina. *Proc. Honey Bee Tracheal Mite (Acarapis woodi R.) Sci. Symp., Am. Assoc. Prof. Apic. and Can. Assoc. Prof. Apic.,* 50.

Nasr, M.E., R.W. Thorp, T.L. Tyler & D.L. Briggs. 1990. Estimating honey bee (Hymenoptera: Apidae) colony strength by a simple method: measuring cluster size. *J. Econ. Entomol.* 83 (3), 748-754.

Nelson, D.L., G. Grant & D. McKenna. 1991. Comparison of colony with cage trials for evaluating queenlines for resistance to honey bee tracheal mites. Proc. Amer. Bee Res. Conf., *Am. Bee J.* 131 (12), 779.

New Treatment Studied to Combat Spreading Tracheal Bee Mite. 1989. USDA Office of Information. *Am. Bee J.* 129 (10), 673-674.

Nixon M. 1982. Preliminary world maps of honeybee diseases and parasites. *Bee Wld.* 63 (1), 23-42.

Örösi-Pál, Z. 1943. Experiments on the feeding habits of the Acarapis mites. *Bee Wld.* 15, 93-94.

Otis, G. W. & C. D. Scott-Dupree. 1992. Effects of *Acarapis woodi* on overwintered colonies of honey bees (Hymenoptera: Apidae) in New York. J. Econ. Entomol. 85 (1): 40-46.

Otis, G. W. 1990. Results of a survey on the economic impact of tracheal mites. Am. Bee J. 130 (1): 28-41.

Otis, G.W., G.M. Grant, D.L. Randall, & J.B. Bath. 1986. Summary of the tracheal mite project in New York. *Proc. Honey Bee Tracheal Mite (Acarapis woodi R.) Sci. Symp., Am. Assoc. Prof. Apic. and Can. Assoc. Prof. Apic.,* 22-30.

Otis, G.W., J.B. Bath., D. L. Randall & G.M. Grant. 1988. Studies of the honey bee tracheal mite (*Acarapis woodi*) (Acari: Tarsonemidae) during winter. *Can. J. Zoo.* 66, 2122-2127.

Page, Jr. R.E. & N. Gary. 1989. Genotypic variation in susceptibility of honey bees, *Apis mellifera,* to infestation by tracheal mites, *Acarapis woodi. Exp. Applied Acarlo.* in press

Peng, Y.S. and M.E. Nasr. 1985. Detection of honeybee tracheal mites (*Acarapis woodi*) by simple staining techniques. *J. Invertebr. Pathol.* 46 (3), 325-331.

Pettis J. S. & W. T. Wilson. 1995. Life history of the honey bee tracheal mite (Acari: Tarsonemidae). *Ann. Ento. Soc. Am..* 89: 368-374.

Pettis, J. S. & W.T. Wilson. 1990. Life cycle comparisons between *Varroa jacobsoni* and *Acarapis woodi.* Am. Bee J. 130 (9): 597-599.

Pettis, J. S., W.T. Wilson & F. A. Eischen. 1992. Nocturnal dispersal by fecund *Acarapis woodi* in honey bee (*Apis mellifera*) colonies. Exp. Appl. Acarol. 15: 99-108.

Pettis, J.S., A. Dietz and F. A. Eischen. 1989. Incidence rates of *Acarapis woodi* (Rennie) in queen honey bees of various ages. *Apidologie* 20 (1), 69-75.

Pettis, J.S., R.L. Cox & W.T. Wilson. 1988. Efficacy of fluvalinate against the honey bee tracheal mite, *Acarapis woodi,* under laboratory conditions. Proc. 3rd. Amer. Bee Res. Conf. *Am. Bee J.* 128, 806.

Phelan, L. P., A.W. Smith & G. R. Needham. 1991. Mediation of host selection by cuticular hydrocarbons in the honey bee tracheal mite *Acarapis woodi* (Rennie). J. Chem. Ecology. 17(2): 463-473.

Phillips, E. F. 1922. The occurrence of disease of adult bees. USDA Circular #218. Washington, D.C. 16pp.

Phillips, E. F. 1923. The occurrence of disease of adult bees II. USDA Circular #287. Washington, D.C. 30pp.

Proceedings of the Third American Bee Research Conference. 1988. *Am. Bee J.* 128 (12), 799-812.

Ragsdale, D.W. & B. Furgala. 1987. A serological approach to the detection of *Acarapis woodi* parasitism in honey bees using an enzyme-linked immunosorbent assay. *Apidologie* 18 (1), 1-10.

Ragsdale, D.W. & K.M. Kjer. 1989. Diagnosis of tracheal mite (*Acarapis woodi* Rennie) parasitism of honey bees using monoclonal based enzyme-linked immunosorbent assay. Am. Bee J. 129 (8), 550-553.

Rennie, J. 1923. Acarine disease explained. North Scotland Coll. Agric. Bull. No. 6. 50pp.

Rennie, J., P.B. White & E.J. Harvey. 1921. Isle of Wight disease in hive bees. *Trans. Royal Soc. Edinb.* 52, 737-739.

Robinson, F.A., K.L. Thel, R.C. Littell & S. B. Linda. 1986. Sampling apiaries for honey bee tracheal mite (*Acarapis woodi* Rennie): Effects of bee age and colony infestation. Am. Bee J. 126, 193-195.

Royce, L. A. & P.A. Rossignol. 1990. Epidemiology of honey bee parasites. *Parasitology Today* 6: 348-353.

Royce, L.A. & P.A. Rossignol. 1990. Honey bee mortality due to tracheal mite parasitism. *Parasitology* 100 (1), 147-151.

Royce, L.A. & P.A. Rossignol. 1991. Sex bias in tracheal mite [*Acarapis woodi* (Rennie)] infestation of honey bees (*Apis mellifera* L.). BeeScience. 1 (3): 159-161.

Royce, L.A., G.W. Krantz, L.A. Ibay, & D. M. Burgett. 1988. Some observations on the biology and behavior of *Acarapis woodi* and *Acarapis dorsalis* in Oregon, In: *Africanized Honey Bees and Bee Mites* (Ed. by G. R. Needham, R. E. Page, Jr., M. Delfinado-Baker, & C. E. Bowman), pp.498-505. Chichester: Ellis Horwood, Ltd.

Royce, L.A., P.A. Rossignol, D. M. Burgett & B.A. Stringer. 1991. Reduction of tracheal mite parasitism of honey bees by swarming. *Phil. Trans. R. Soc. Lond. B*, 331, 123-129.

Sachs, H. 1951. Zur morphologie von *Acarapis woodi* . I. Bau und funktion der mundwerkzeuge der Tracheenmilbe, *Acarapis woodi* (Rennie), 1921. [On the morphology of *Acarapis woodi* . I. Structure and function of mouth-parts of the tracheal mite]. Z. Bienenforsch. 1: 103-112.

Sachs, H. 1953. Zur morphologie von *Acarapis* . I. Bau und funktion der Biene von *Acarapis woodi* . [Structure and function of the bee mite *Acarapis woodi*]. Z. Bienenforsch. 2: 1-7.

Sachs, H. 1958. Versuch zur Züchtung der Tracheenmilbe, *Acarapis woodi woodi* . [Experiments on rearing the tracheal mite *Acarapis woodi woodi*]. Z. Bienenforsch. 4: 107-113.

Sammataro, D. 1995. Studies on the control, behavior and molecular markers of the tracheal mite (Acarapis woodi [Rennie]) of honey bees (Hymenoptera: Apidae). Ph.D. dissertation. The Ohio State University, Columbus, OH.

Sammataro, D., P.G. Parker & G. R. Needham. 1995. Using PCR-based RAPDs (Random amplified polymorphic DNA) to determine differences in tracheal mite *Acarapis woodi* (Rennie) (Acari: Tarsonemidae) populations. Proceedings IX Acarology Congress, Columbus OH.

Sammataro, D. 1990. Tracking tracheal mites. *Gleanings in Bee Cult.* 118 (4), 206-208.

Sammataro, D. & G.R. Needham. Host-seeking behaviour of tracheal mites (Acari: Tarsonemida) on honey bees (Hymenoptera: Apidae). *Exp. Appl. Acarol.* 20: 121-136. 1996.

Sammataro, D. & G.R. Needham. 1996. How oil affects the behavior of tracheal mites. *ABJ*, 136: 511-514.

Sammataro, D. 1997. Report on parasitic honey bee mites and disease associations. *ABJ*, 137: 301-302.

Sammataro, D., S. Cobey, B.H. Smith & G.R. Needham. 1994. Controlling tracheal mites (Acari: Tarsonemidae) in honey bees (Hymenoptera: Apidae) with vegetable oil. *J. Econ. Entomol.* 87: 910-916.

Shimanuki, H. & D. Knox. 1991. *Diagnosis of honey bee diseases.* USDA Agri. Handbook No. AH-690.

Shimanuki, H. and D.A. Knox. 1989. Tracheal mite surveys. *Am. Bee J.* 129 (10), 671-672.

Shimanuki, H., D.A. Knox, M. Delfinado-Baker and P.J. Lima. 1983. National honey bee mite survey (1). *Apidologie*, 14, 329-332.

Smith, A. W. 1991. Population dynamics and chemical ecology of the honey bee tracheal mite *Acarapis woodi* (Acari: Tarsonemidae). Ph.D. dissertation. The Ohio State University, Columbus, OH.

Smith, A.W. & G. R. Needham. 1988. A new technique for the rapid removal of tracheal mites from honey bees for biological studies and diagnosis, pp. 530-534. *In* Needham, G.R., R.E. Page, Jr., M. Delfinado-Baker, & C.E. Bowman eds. *African Honey Bees and Bee Mites.* Ellis Horwood, Ltd. Chichester.

Smith, A.W. & G.R. Needham. 1989. Honey-bee tracheal mite *Acarapis woodi* Rennie. Fact Sheet. Acarology Dept. The Ohio State Un. Columbus.

Smith, A.W., G.R. Needham & R.E. Page, Jr. 1987. A method for the detection and study of live honey bee tracheal mites (*Acarapis woodi* Rennie). *Am. Bee J.* 127 (6), 433-434.

Smith, A.W., G.R. Needham, R.E. Page, Jr. & M. Kim Fondrk. 1991. Dispersal of the honey-bee tracheal mite, *Acarapis woodi* (Acari: Tarsonemidae) to old winter bees. *Bee Science* 1 (2), 95-99.

Smith, A.W., R. E. Page, Jr., & G. R. Needham. 1991. Vegetable oil disrupts the dispersal of tracheal mites, *Acarapis woodi* (Rennie), to young host bees. *Am. Bee J.* 131 (1): 44-46.

Sturges, A. M. 1921. Observations on acarine disease. *Bee World* 3: 188-190.

Sugden, E. A., K. R. Williams & D. Sammataro. IXth International Congress of Acarology: a honey bee mite round table. *Bee Culture*, 123 (2): 80-81. 1995.

Thoenes, S. C. & S. L. Buchmann. 1992. Colony abandonment by adult honey bees: a behavioral response to high tracheal mites infestation? *J. Apic. Res.* 31(3/4): 167-168.

Thoenes, S.C. and S.L. Buchmann. 1990. Tracheal mite induced colony mortality monitoring electronic scale. *Am. Bee J.* 130, 816.

Tomasko, M., J. Finley, W. Harkness & E. Rajotte. 1993. A sequential sampling scheme for detecting the presence of tracheal mite (*Acarapis woodi*) infestations in honey bee (*Apis mellifera* L.) colonies. *Bulletin No. 871*, Penn State Ag. Exper. Sta., University Park PA.

Vecchi, M.A. and G. Giordani. 1986. Chemotherapy of acarine disease. 1. Laboratory tests. *J. Invertebr. Pathol.* 10, 390-416.

Waller, G.D. and L.H. Hines. 1990. A search for tracheal mite resistance in Arizona honey bees. *Am. Bee J.* 130, 818.

Wehrle, L. P. and P. S. Welch. 1925. The occurrence of mites in the tracheal system of certain Orthoptera. Ann. Entomol. Soc. Amer. 18 (1): 35-44.

Wille, H., A. Geiger & A. Muff. 1987. Mitteilungen der sektion bienen: Einfluss der milbe *Acarapis woodi* auf den Massenwechsel von Bienenvölkern. Schweizerische Bienenzeitung, 110, 346-348, CH-3097 Liebefeld-Bern, Switzerland.

Wilson, W.T. 1990. Tracheal mites (*Acarapis woodi*): an overview. Abstract, Eastern U.S. Tracheal Mites Symposium. *Am. Bee J.* 130, 185.

Wilson, W.T. & A. M. Collins. 1991. Spring applications of formic acid for control of *Acarapis woodi.* Proc. Amer. Bee Res. Conf., *Am. Bee J.* 131 (12),785.

Wilson, W.T., Cox, R.L. and Moffett J.O. 1990. Menthol-grease board: a new method of administering menthol to honey bee colonies. *Am. Bee J.* 130 (6), 409-412.

Wilson, W.T., J. O. Moffett, R.L. Cox, D.L. Maki, H. Richardson & R. Rivera. 1988. Menthol treatment for *Acarapis woodi* control in *Apis mellifera* and the resulting residues in honey. In: *Africanized Honey Bees and Bee Mites* (Ed. by G. R. Needham, R. E. Page, Jr., M. Delfinado-Baker, & C. E. Bowman), pp.535-540. Chichester: Ellis Horwood, Ltd.

Wilson, W.T., J. R. Elliott & J. J. Lackett. 1970. Antibiotic treatments last longer. *Am. Bee J.* 110: 348; 351.

Feasibility of apicultural tools for a non-chemical *Varroa* control

JÖRG SCHMIDT-BAILEY[1] AND STEFAN FUCHS[2]

[1]DEPARTMENT OF ENTOMOLOGY RUTGERS, THE STATE UNIVERSITY
93 LIPMAN DRIVE, BLAKE HALL
NEW BRUNSWICK, NJ 08901-8524
joerg@aesop.rutgers.edu

[2]INSTITUT FÜR BIENENKUNDE (POLYTECHNISCHE GESELLSCHAFT)
KARL VON FRISCH WEG 2
D-61440 OBERURSEL, GERMANY

Summary

The integration of biological *Varroa* control by drone brood trapping combs (DTC) into a swarm prevention method was tested in fifteen *Apis mellifera* colonies between May and August 1996 in Germany. The fifteen colonies were divided into ten experimental colonies and five control colonies. All colonies became temporarily broodless during the swarm prevention by colony-splitting. During this broodless time drone brood combs (produced within the colonies) were inserted into the ten experimental colonies and became therefore the only remaining brood for *Varroa* mites to invade. Each colony had to produce three drone brood combs for the use of trapping combs. All trapping combs were removed immediately after being sealed and stored in a freezer. Later all cells were uncapped and the numbers of invaded mites were counted. The remaining mites in the experimental colonies and the mites in the control colonies were killed with Bayvarol®. All split colonies were reunited at the end of the experiment.

Results: The number of mites in the trapping combs (sum of all three trapping combs) was 963 ± 362. The number of the remaining mites in the experimental colonies was 49 ± 22. The efficiency of the control method (sum of all three trapping combs) was $94.9 \pm 2\%$. The number of mites in the control colonies was 1393 ± 304.

Introduction

The persistent threat of *Apis mellifera* colonies by *Varroa jacobsoni* requires annual control procedures. Using chemical treatments (pesticides) has the positive aspects of high efficiency and easy application, but there are several significant negative aspects. The fat soluble pesticides (pyrethroids like Apistan®) contaminate beeswax and honey (Wallner 1994) and can lead to resistant mites (Milani 1995). The water soluble pesticides (organic acids like formic acid) can also contaminate honey if applied during a honeyflow. Both kinds of pesticides should therefore not be applied during a honeyflow. During this period the mite population is increasing to a level of possibly damaging emerging bees (De Jong *et al.* 1982), the risk of secondary infections like virus diseases (Ball 1983) and a further spreading of mites to other colonies by robbery (Sakofski 1988). To prevent not only the risk of contaminated beeswax and honey but also the build up of a damaging mite level, an alternative control method must be developed. This method should remove the major part of a mite population during the honeyflow without using any pesticides and should be also integrated into common apicultural procedures to avoid additional work. A possible alternative method occured in the research results of Boot *et al.* (1995), who estimated that only a small amount of drone brood is needed to trap a major amount of mites in broodless colonies. When applying swarm prevention methods parts of colonies become temporarily broodless and this fact was used to combine swarm prevention and mite control (Calis *et al.* 1996; Schmidt-Bailey *et al.* 1996). These experiments showed that a single drone brood

KEY:

H	HONEY CHAMBER
B	BROOD CHAMBER
DTC	DRONE BROOD TRAPPING COMB
YELLOW COMB	DRONE CELL FOUNDATION
YELLOW/WHITE COMB	DRONE COMB CONTAINING EGGS
YELLOW/BROWN COMB	SEALED DRONE BROOD COMB
GRAY HIVE PARTS	ALREADY IN USE IN MAY
RED HIVE PARTS	ADDITIONAL HIVE PARTS
♀	ORIGINAL QUEEN
♀̄	QUEEN REARING

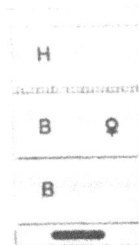

Figure 1. Condition of an *Apis mellifera* colony in May. Two brood chambers and the first honey super are seperated by a queen excluder. (Ten 16x8 frames per chamber and super).

trapping comb is able to trap 88% of a mite population in a temporary broodless colony. With two drone brood trapping combs in a row, the trapping efficiency is raised up to 98%. The following experiment was a critical test of the existing and very promising results.

Methods and Materials

In May, 1996 fifteen *Apis mellifera* colonies, each containing two brood chambers, a queen excluder and the first honey super (Figure 1) were divided into five control colonies and ten experimental colonies.

Control Colonies. At the beginning of the experiment a second queen excluder was inserted into each control colony to separate the two brood chambers (Figure 2).

One week later the honey super was removed, and it was determined which of the two brood chambers contained eggs and larvae (queen!). The colonies were split into a temporary colony containing the brood chamber with the queen and the remaining queenless colony containing the other brood chamber and the honey super. The temporary colony was put a few meters aside while the re-

maining colony stayed at the old spot (Figure 3). In the remaining colony (only sealed brood!) an artificial queen cell containing a young larva was inserted to produce a new queen. The temporary colony and the remaining colony were separated until July and then reunited. To determine the number of mites, a Bayvarol® treatment was carried out subsequently.

Experimental Colonies. In each experimental colony two frames with drone cell foundation (one per brood chamber) and a second queen excluder were inserted (Figure 4).

A week later the honey super was removed. Then it was determined which of the two brood chambers contained the drone brood comb with eggs and larvae (queen). The colonies were split into an artificial swarm and a queenless brood sampler (Figure 5). The artificial swarm contained the drone brood comb with eggs (trapping comb 1), no other brood, a few foundations, all bees and the queen out of one brood chamber, a queen excluder, the honey super and stayed on the original location. The brood sampler contained all brood combs and was put a few meters to the side.

One week after splitting, the trapping comb 1 in the artificial swarm was completely sealed and could be removed and replaced with an empty comb or foundation. In the brood sampler all emergency queen cells were destroyed and an artificial queen cell containing a young larva was inserted.

Two weeks after splitting, a drone cell foundation was inserted into the artificial swarm (to deliver trapping comb 2 the following week). The now sealed queen cell was replaced with a new one containing a young larva (the replacement should guarantee a longer broodless period later on in the experiment).

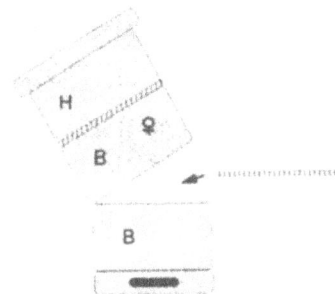

Figure 2. Separating brood chambers in a control colony by inserting a second queen excluder.

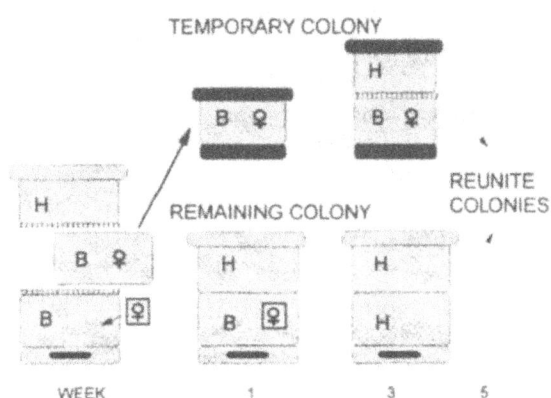

Figure 3. Splitting a control colony within a swarm prevention routine. There is no time consumed searching for the queen. Simply find eggs ar larvae in one of the brood chambers and put the complete chamber aside.

Three weeks after splitting, the trapping comb 2 (containing eggs and larvae) was removed from the artificial swarm, replaced by a drone cell foundation to deliver trapping comb 3 the following week and inserted into the brood sampler where all the bees and with them all the mites had emerged in the meantime. The trapping comb was the only possibility for mites to invade any brood cells.

Four weeks after splitting, the trapping comb 2 in the brood sampler was completely sealed and was removed and replaced by trapping comb 3. Because the young queen did not lay eggs, the trapping comb was the only possibility for the remaining mites to invade brood cells.

Five weeks after splitting, the trapping comb 3 in the brood sampler was completely sealed and removed. The artificial swarm and brood sampler were then treated with Bayvarol® to determine the number of remaining mites and reunited subsequently. The harvest of honey in the experiment took place before the colonies were treated with Bayvarol®.

Results

The number of mites in the control colonies (n= 5) was 1393 ± 304 at the end of the experimental period (Table 1). The number of mites trapped in drone brood combs (sum of all three trapping combs) was 963 ± 362. The number of remaining mites in the experimental colonies after trapping comb treatment was 49 ± 22. The efficiency of the control method (sum of all three trapping combs per colony) was 94.9 ± 2 %. All ten experimental

colonies had produced the necessary amount of three drone brood combs within the experimental period. The honey yield in the control colonies was 29.8 ± 2.5 kg compared with 31 ± 5 kg in the experimental colonies.

Discussion

The experiment showed that *Apis mellifera* colonies can produce a sufficient number of drone brood combs. In the experiment the combs were produced out of drone cell foundation. Another possibility could be to insert frames with only one inch of foundation at the top. Strong colonies will fill such a frame with drone cells as well, especially in the beginning of the swarm season.

The use of swarm prevention by splitting colonies is a common procedure by which the number of colonies is doubled. After the swarm season both parts of the original colony have turned into full comparable colonies. This is a big advantage because it can then be decided whether to reunify the colonies to keep the original amount or to leave them doubled. The observed 94% efficiency of the drone brood trapping comb method in this experiment conforms with the results of Calis *et al.*(1996) and Schmidt-Bailey *et al.*(1996). This high efficiency decreased the mite population to an innocuous level as early as June/July. This is very important for the development of a healthy population of wintering bees. But it is necessary to consider that there is brood in bee colonies until August/September. The remaining mites will invade this brood and reproduce. Thus the mite population will increase from under fifty mites in July to a few hundred in

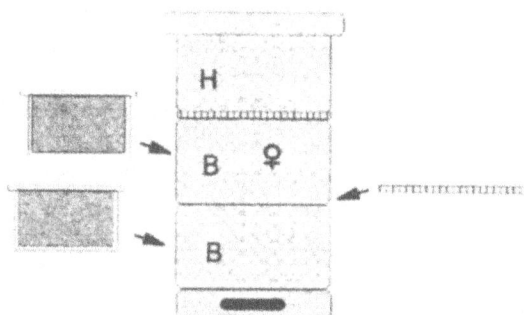

Figure 4. Separating brood chambers in an experimental colony by inserting a second queen excluder. The simultaneous inserting of a drone cell foundation per brood chamber will deliver the first trapping comb a week later.

September. This number will not be a threat for the colony, but it requires a repeat of the mite control with the same accuracy in the following season. The three necessary trapping combs per colony produce an excess of sealed drone brood in proportion to the number of treated colonies. The easiest way to handle all these combs is a wax melter. An alternative is decapping the cells and removing the larvae and mites with a jet of water. These cleaned combs can be stored and reused in the following season.

Conclusions

The experiment demonstrated the possibility of combining swarm prevention, *Varroa* control, queen rearing, honey harvesting and doubling of colonies (or reunification) in a convenient weekly schedule. It demonstrated also that it is conclusively possible to produce the needed amount of drone brood in *Apis mellifera* colonies to be used as trapping combs during swarm control within these colonies. The efficiency of only three drone brood trapping combs, inserted in temporary broodless colonies, is almost 95% .

Therefore, every beekeeper can decrease the quantity of mites to an innocuous level in his routine apicultural procedures without the use of chemicals.

Acknowledgments

We are very grateful for the help from colleagues at Institut fuer Bienenkunde (Polytechnische

	TRAPPED MITES	REMAINING MITES	TRAPPING EFFICIENCY %	HONEY kg
CONTROL COLONIES				
1		1599		33
2		1202		29
3		1417		26
4		938		32
5		1813		29
Ø		1393±304		29,8 ± 2,5
EXPERIMENTAL COLONIES				
1	1472	33	97,8	25
2	560	52	91,5	31
3	952	34	96,6	33
4	1257	64	95,2	34
5	599	12	98	22
6	966	69	93,3	27
7	670	40	94,4	32
8	1340	64	95,4	39
9	1372	93	94	38
10	442	33	93,1	29
Ø	963±362	49±22	94,9 ± 2	31±5

Table 1. Single results and medians for trapped varroa (trapping combs 1+2+3 added), remaining varroa (killed with Bayvarol®), trapping efficiency (Trapping combs 1+2+3 added) and honey yield.

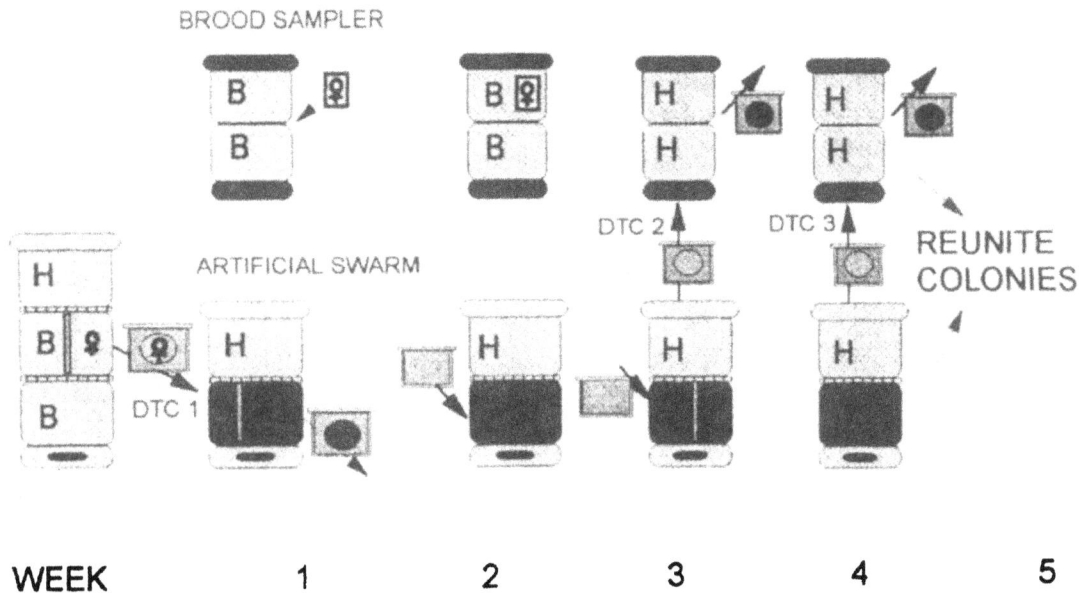

WEEK   1   2   3   4   5

Figure 5. Combination of swarm prevention and mite control. There is no consumed searching for the queen. Simply find the drone brood comb containing eggs and brush the whole brood chamber as an artificial swarm.

Schmidt-Bailey & Fuchs: Non-chemical varroa control

Chapter 14

Gesellschaft) Oberursel Germany, Hessische
Landesanstalt für Tierzucht Abteilung Bienenzucht
Kirchhain Germany and Department of Entomology Agricultural University Wageningen The Neth-
erlands.

References

Ball, B.V. 1983. Der Zusammenhang zwischen *Varroa
    jacobsoni* und Viruserkrankungen der Honigbiene.
    Allgemeine Deutsche Imkerzeitung 17: 177-143

Boot, W.J., J. Schoenmaker, J.N.M. Calis and J.Beetsma
    1995. Invasion of Varroa mites into drone brood cells
    of the honey bee. *Apidologie* 26: 109-118

Calis, J.N.M., W.J. Boot, J. Beetsma, J.H.P.M. van den
    Eijnde, A. de Ruijter and J.J.M. van der Steen, 1996.
    Trapping *Varroa jacobsoni* drone brood combs of *Apis
    mellifera* in broodless colonies. *Apidologie* 27 (4): 295.

De Jong, D., P.H. De Jong and L.S. Goncalves 1982.
    Weight loss and other damage to developing worker
    honeybees from infestation with *Varroa jacobsoni. J.
    Apic. res.* 21: 165-165

Milani, N. 1995. The resistance of *Varroa jacobsoni* to
    pyrethroids: A laboratory assay. *Apidologie* 26: 361-
    364

Sakofski, F. 1988. Transfer of *Varroa jacobsoni* by rob-
    bing. Proceedings of a meeting of the EC-Experts
    Group, Udine, Italy, 28 to 30 November 1988, Present
    status of varroatosis in Europe and progress in the
    Varroa-mite control, R. Cavalloro, 1989: 177-181

Schmidt-Bailey, J., S. Fuchs and R. Büchler 1996. Effec-
    tiveness of drone brood trapping combs in broodless
    honeybee colonies. *Apidologie* 27 (4): 294.

Wallner, K. 1994. Zur Rueckstandsfrage in der
    Varroabehandlung (Residues in honey and beeswax
    caused by *Varroa* treatment. *Apidologie* 25(5): 505-506

**Keywords: biological *Varroa* control, drone brood
trapping combs, swarm prevention, *Apis
mellifera*, *Varroa jacobsoni***

Resistance to *Varroa jacobsoni* by *Apis mellifera*: A perspective for the beekeeping industry

MARLA SPIVAK AND OTTO BOECKING

**DEPARTMENT OF ENTOMOLOGY, 219 HODSON HALL
UNIVERSITY OF MINNESOTA, ST. PAUL, MN 55108, USA**

**INSTITUT FÜR LANDWIRTSCHAFTLICHE ZOOLOGIE UND BIENENKUNDE
DER UNIVERSITÄT, MELBWEG 42, 53127 BONN, GERMANY**

Introduction

Beekeeping with *Apis mellifera* L. is endangered worldwide by the parasitic mite *Varroa jacobsoni* Oud. *A. mellifera* colonies die from varroa infestation within a few years if the mite population is not regulated by the beekeeper. Chemical treatments are used extensively in honey bee colonies to reduce the damage caused by the varroa mite, but the use of miticides is controversial because of the risk of contaminating bee products with chemical residues and the development of resistant mites (see Milani, these proceedings). Because of the disadvantages of using chemical treatments in mite infested colonies, it is of common interest to select for colonies that are able to survive mite infestations without regular treatment. Such colonies would display resistance; they would have defenses against the mite rather than succumb to parasitism.

For colonies to survive any extended period of time without treatment, the survivorship and/or reproductive success of the mites must be reduced. The population growth of varroa mites is influenced by the following factors (Fries *et al.* 1994, Harbo 1996): 1) the successful entry of a mated, female mite into a brood cell containing a fifth instar larva; 2) the successful reproduction of the mite within the cell; 3) the probability that the same mite will survive to enter another brood cell; and 4) the number of complete reproductive cycles a mite can complete within a season.

In an obligate host-parasite interaction such as that between varroa and the honey bee, it is important to differentiate between the effects of genetics and of the environment on both the bee and the mite. Considerable attention has been directed toward the apparent resistance of different genetic races of bees, but until recently, little attention has been directed to the genetics of the mite. Is *Varroa jacobsoni* the only species of varroa found on *A. mellifera* worldwide? Are there varieties of mites that are more or less virulent? Is mite fertility a function of mite genetics and physiology or bee genetics and physiology? Environmental effects such as resource abundance and quality, and the duration and intensity of winter, have direct effects on the survival and population growth of honey bee colonies, but do they also affect the mites? Environmental effects on bees or mites may temporarily influence the reproductive success of the mites, but will not lead to sustainable resistance. Only bees that have genetic mechanisms of increasing the mortality and/or decreasing the fertility of the mites will demonstrate some degree of resistance.

This paper will be divided into the following sections: 1) mite genetics; 2) bee genetics; 3) bee environment; 4) mite environment; and 5) current breeding programs. It is in no way intended to be a thorough review of all the research that has been done on these topics. Our aim is to highlight some of the key research findings in each topic to help focus attention on areas where more research is needed. An updated version of this chapter will be found in Spivak and Boecking (in press). We also recommend other review articles with somewhat different emphases (Boecking and Ritter 1994; Büchler 1994; Fries *et al.* 1994; Boecking and Spivak, in press).

I. Mite genetics

Recent observations (Anderson 1994; Anderson and Sukarsih 1996; Anderson and Fuchs 1998) of varroa infestation on *Apis cerana* Fabr. and *Apis mellifera* in Papua New Guinea and islands in Indonesia (Java, Irian Jaya, Biak and Yapen) have yielded some interesting new insights into the bee-mite interaction. Varroa mites reproduce successfully on drone brood but not on worker brood of *A. cerana*, which is typical of varroa infestations on this species of honey bee throughout most of its distribution in Asia. During 1992-1995 in Papua New Guinea and Irian Jaya, Anderson observed that although the *A. mellifera* colonies in the same areas were infested with varroa, the mites were not reproducing on either the drone or worker brood. The lack of reproduction was observed also in Java during 1991 and 1992, but in subsequent years (1993-1995), mites were found reproducing in sealed worker and drone brood of *A. mellifera*. Anderson and Sukarsih (1996) speculated that the observed change in varroa reproductive behavior in *A. mellifera* colonies in Java may be due to a new "strain" of varroa introduced to Java after January 1992. Recent experiments and DNA analyses of the mite indicates that in fact there now may be two types of varroa in this area (Anderson and Fuchs 1998).

Other research on molecular genetic differences among mites from different locations has been conducted by de Guzman *et al.* (1996). They concluded that the varroa of North America is Russian in origin *via* Europe, while mites of Brazil and Puerto Rica may be Japanese in origin. Current research is underway to determine the extent of these genetic differences among mites from different areas. Continued research is necessary to determine whether the genetic differences are related to differing reproductive success of the mites on the bee populations where they are found.

II. Bee genetics

Certain races and lines of bees appear to have genetic mechanisms which enable them to resist

GROOMING BEHAVIOR

AUTO-GROOMING   ALLO-GROOMING

Figure 1. Some honey bees are able to remove mites from themselves (auto-grooming), or from their nestmates (allo-grooming). Drawing by O. Boecking.

the mites, although there is limited information on the nature of the mechanisms. There are reports that indicate the possibility of varroa resistance in Africanized bees from Brazil (De Jong *et al.* 1984; Camazine 1986), European honey bees from Italy (Monaco 1997) and Uruguay (Ruttner *et al.* 1984, and *A. mellifera intermissa* from Tunisia (Ritter 1990; Ritter *et al.* 1990; Ritter 1993). In these areas, varroa is reported to have low reproductive success on worker brood. Although some *A. m. intermissa* colonies in Tunisia have survived high varroa infestations without treatment for several years, the low reproductive success of varroa on worker brood does not seem to be a stable phenomenon (Ritter and Boecking, unpublished observations). In Uruguay, no treatment for varroa has been necessary over the last 15 years. However, recent investigations could not substantiate previous observations by Ruttner (1984) that the resistance was due to a high percentage of infertile mites. European queens from Uruguay, purported to be resistant, were imported into France and Poland (N. Koeniger, personal communication). There, the number of reproductive mites was the same as in susceptible colonies from Europe. The colonies with queens from Uruguay had larger brood areas in Europe, which stimulated the growth of mite populations.

The situation in Brazil, however, continues to be intriguing. The reproductive success of varroa on worker brood of Africanized bees appears to be much lower than on worker brood of European bees situated in the same area (Camazine 1986; Rosenkranz and Engels 1994). Chemical treatments of infested Africanized colonies apparently are not necessary in Brazil.

In Mexico, where Africanized bees have been present for a much shorter time than in Brazil, tests revealed that European brood was twice as attractive to varroa than Africanized brood. However, the reproductive success of varroa was the same in worker brood of Africanized and European colonies and was significantly higher on worker brood of reciprocal F_1 hybrids (Guzman-Novoa et al. 1996).

Continued research is necessary to determine whether the differences in fertility are due to the genetics of the mite or the bee, and the extent that the reproductive differences are influenced by environmental conditions and colony characteristics. The low reproductive success of mites in worker brood compared to the high reproductive success in drone brood may alone explain varroa resistance in Africanized A. mellifera colonies in Brazil (Fuchs 1992). In addition, the intense tendency to swarm and the high density of feral colonies may contribute to the rapid development of natural resistance (Engels et al. 1986). Defense mechanisms, such as grooming and brood removal behavior (see below), may provide additional explanations for resistance of A. mellifera colonies in Brazil and Tunisia (Moretto et al. 1991a 1993; Boecking and Ritter 1993).

Bee post-capping duration

An example of a heritable physiological mechanism that may affect the reproductive rate and population growth of the mite is the duration of the post-capping, or pre-pupal and pupal stages of the bees. The Cape bee from South Africa, A. m. capensis has a shorter duration of the post-capping stage (pupal stage), which may result in lower mite population growth in these colonies (Moritz and Hänel 1984). Although others (Moritz 1985; Schousboe 1986; Le Conte and Cornuet 1989; Büchler and Drescher 1990; Moritz and Mautz 1990) have speculated that post-capping duration is a significant factor in limiting mite reproduction, at present, no one has provided data from A. mellifera that supports the hypothesis that there is a correlation between the duration of the capped bee brood stage and the reproduction success rates of the mite.

Bee behavioral defenses

Two genetically determined behaviors of honey bees, grooming behavior (Figure 1) and removal (hygienic) behavior (Figure 2), may also limit the survivorship and reproductive success of varroa.

Grooming. In grooming behavior, an adult bee removes the mites from herself (auto-grooming), or from infested nestmates (allo-grooming) (Peng et al. 1987a; Rath 1991; Büchler et al. 1992; Takeuchi 1993; Fries et al. 1996). The survivorship of the mite decreases if the bee successfully damages the mite with her mouthparts during the grooming process. If the mites are simply disturbed during grooming and fall to the bottom of the colony without being damaged, they are able to subsequently reproduce.

The first study of grooming behavior was conducted by Peng et al. (1987a) in China. Observation hives containing A. cerana and A. mellifera were inoculated with mites collected from A. mellifera colonies. 98% of the introduced mites were either removed by the bees by auto- or allo-grooming or moved onto another bee in the A. cerana colonies, while only 3% of the mites were groomed or moved to another bee in the A. mellifera colonies. Other researchers have confirmed that A. mellifera does exhibit grooming behavior, but to a lesser extent than A. cerana (Wallner 1990; Rath 1991; Büchler et al. 1992; Moosbeckhofer 1992; Ruttner and Hänel 1992; Boecking and Ritter 1993; Moretto et al. 1993). Recent investigations by Fries et al. (1996) indicate that grooming behavior may be less effective in combating the mites than previously reported. When mites were introduced into field colonies of A. cerana and A. mellifera, the percent of live mites with visible damage as a consequence of grooming was about 30% in A. cerana compared to 12.3% in A. mellifera colonies. In observation hives, 12.5% of the recovered mites were damaged by A cerana and none were damaged by A. mellifera. Fries et al. (1996) tabulated only incidents of successful grooming that resulted in damage to the mite, in contrast to Peng et al. (1987a) and Büchler et al. (1992), who included the movement of mites from one bee to another bee and the loss of mites from the observation area by the observer as grooming events.

Although the proportion of damaged mites may vary widely from colony to colony, the overall effect of grooming behavior on the reduction of the mites within bee colonies remains unclear.

Removal Behavior. In removal, or hygienic, behavior, an adult bee detects, uncaps, and removes an infested pupa from the cell (Peng et al. 1987b; Rath and Drescher 1990; Rosenkranz and Tewarson 1992; Tewarson et al. 1992). The removal of infested

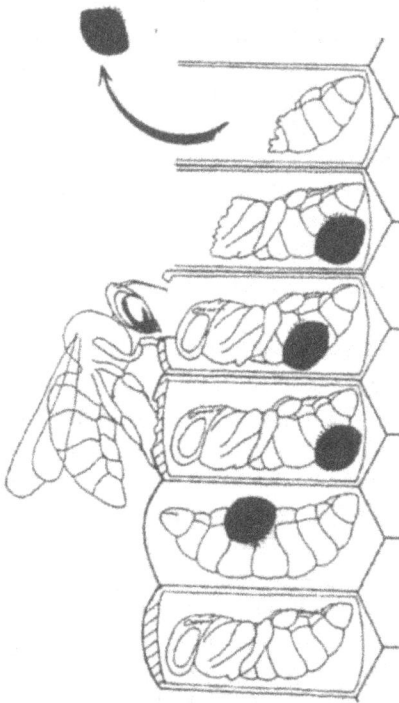

Figure 2. The process of removal, or hygienic, behavior. A bee detects a cell containing a parasitized pupa. She uncaps the cell, and the pupa is removed from the cell. The mother mite may escape from the cell, but her offspring would be destroyed by the bees during the removal process. Drawing by O. Boecking.

pupae may theoretically limit the growth of a varroa population in three ways: 1) immature mites which have begun to develop in brood cells are killed, decreasing the average number of offspring per mother mite; 2) the mother mites may be damaged during the removal process, and 3) the phoretic period of a mother mite is extended if she escapes during the removal process (Fries *et al.* 1994).

Peng *et al.* (1987b) were also the first to describe removal behavior in colonies of *A. cerana* in China. A significant correlation was found between the infestation level of the observation hives containing *A. cerana* or *A. mellifera* and the degree of removal of infested pupae by the bees. Subsequently, Rath and Drescher (1990) found that *A. cerana* detected and removed 98.8% of the worker pupae experimentally infested with varroa. Investigations by Rosenkranz *et al.* (1993a) indicated that the amount of brood removed by *A. cerana* after experimental infestation with live mites may be lower than previously reported (Peng *et al.* 1987b; Rath and Drescher 1990) because the source of mites

used for experimental infestation influences the removal behavior of the bees.

Because varroa has limited or no reproduction in worker cells of *A. cerana*, the successful reproduction of the mite is limited to its seasonal occurrence of drone brood (Koeniger *et al.* 1981 1983; Tewarson 1987; Rath 1991, 1993; Tewarson *et al.* 1992). *A. cerana* workers do not remove mite-infested drone brood due to the thick cell capping over the drone cell, a structure unique to this species (Rath and Drescher 1990; Rath 1991). Drones which are infested with multiple mites become weakened and are not able to open their cell caps and they die together with the mites inside such cells (Koeniger *et al.* 1983; Rath 1991, 1992). The combination of non-reproduction in worker cells and the drone cell capping result in low overall rates of parasitism in *A. cerana*.

A. mellifera from North American and European stocks also display removal behavior of varroa mites from capped, infested worker brood cells, but to a limited extent compared to *A. cerana* (Peng *et al.* 1987b; Boecking 1992, 1994; Boecking and Drescher 1990, 1991, 1992, 1993; Boecking and Ritter 1993; Spivak 1996).

In a study in which varroa was experimentally introduced into *A. m. carnica* colonies in Germany, Boecking and Drescher (1992) found that the mean rate of removal by the Carniolan bees on day 10 after infestation was 29.3% when one mite per cell was introduced, and 55.1% when two mites per cell were introduced. A positive correlation (r=0.74) was found between the removal rates towards mite-infested brood and the removal of freeze-killed brood, a commonly used assay to test for hygienic behavior (Boecking and Drescher 1992).

Spivak (1996) used a similar approach in the U.S. to test *A. mellifera* colonies of Italian origin for their ability to remove infested pupae. The colonies tested had been bred *a priori* for two generations for hygienic behavior with the goal of breeding colonies resistant to chalkbrood disease (see Gilliam *et al.* 1983, 1988; and Spivak and Gilliam 1993). The freeze-killed brood assay was used to select colonies for hygienic behavior. Colonies that consistently removed a comb section containing 200 cells of frozen brood within 48 hours were considered hygienic (see procedure in Taber 1987; Spivak and Downey 1998. Daughter queens were raised from the hygienic colonies and were instrumentally inseminated with semen from drones from other hygienic colonies. A non-hygienic line of bees was also bred as a control.

Following the methods of Boecking and Drescher (1991, 1992) one varroa mite per cell was introduced through the removable plugs in the bases of 10-20 cells within a "Jenter Box" containing recently sealed fifth instar larvae within the hygienic and non-hygienic colonies (Spivak 1996). Another group of cells serving as controls had the plugs removed and replaced with no mite introduction. The percent of mite-infested pupae removed by hygienic and non-hygienic colonies 10 days after introduction of the mites was significantly different in both 1994 and 1996 (Figure 3). The hygienic colonies (n=4 and 10, respectively) removed an average of 69.2% and 66.7% of the pupae infested with one mite per cell, while the non-hygienic colonies (n=3 and 6, respectively) removed 10.0% and 22.7% (P <0.01 between bee lines for each year). There also was a significant difference between the removal of infested pupae and controls (P < 0.05) in both years. Tests in 1995 (n=7 hyg, 4 non-hyg) revealed a significant difference only when two mites per cell were introduced (treatment effect: P<0.01) (Spivak 1996).

Experiments concerning the effects of the removal behavior on the chances of survival of single mites published by Boecking (1992) and Boecking and Drescher (1993) revealed that most of the adult female varroa mites (n = 104) which escaped the brood cells after the removal could invade other brood cells (mean = 61.3%). Some mites attached themselves to adult bees (14.6%) and a small percentage of them were killed by the bees (10.9%). The source of mites did not influence the removal response in *A. mellifera* as it did in *A. cerana* (Rosenkranz et al. 1993b) even when mites from *A. cerana* were introduced into *A. mellifera* colonies.

As with grooming behavior, the effect of hygienic behavior on the population regulation of varroa within bee colonies remains unclear. However, because hygienic colonies demonstrate resistance to brood diseases such as American foulbrood (Woodrow and Holst 1942; Rothenbuhler 1964), and chalkbrood (Gilliam et al. 1983, 1988) the trait may be worthwhile to incorporate into honey bee stocks. To determine whether colonies with naturally mated queens from a hygienic line of Italian honey bees would have lower levels of diseases and varroa mites and produce a large honey crop, hygienic colonies were compared to colonies from a commercial line of Italian bees not selected for hygienic behavior

(Spivak and Reuter 1988). In previous studies on the relation between hygienic behavior and resistance to diseases and mites, the test colonies contained instrumentally inseminated queens (Rothenbuhler 1964; Gilliam et al. 1983, 1988; Spivak 1996). This was the first study to evaluate hygienic stock in large field colonies with naturally mated queens. The results of tests conducted in 1995 and 1996 revealed that the hygienic colonies removed significantly more freeze-killed brood than the commercial colonies, had significantly less chalkbrood, had no American foulbrood, and produced significantly more honey than the commercial colonies (Spivak and Reuter 1988). Estimates of the number of varroa mites on adult bees indicated that the hygienic colonies had fewer mites than the commercial colonies in three of four apiaries.

III. Bee environment

There are seasonal and climatic influences on the proportion of non-reproducing varroa in all colonies of *A. mellifera* (Kulincevic et al. 1988; Otten and Fuchs, 1990; Otten, 1991). Some interesting examples of how the environment may affect the bee-mite interaction are found in Brazil on Africanized bees. When varroa-resistant Africanized bees are moved from warmer to cooler climates in Brazil, such as from the lowlands in São Paulo to higher elevations in Santa Catarina, the infestation of varroa on worker brood increases (De Jong et al. 1984; Moretto et al. 1991b). In addition, there is a positive correlation between the amount of pollen stored in the nest and the percent of mites that reproduce on worker brood in lowlands of Brazil (Moretto et al. 1997). Although it is unclear how a cooler climate or the amount of stored pollen could affect the fertility of the mites, these observations emphasize the importance of testing "resistant" bee stocks in different environmental conditions to ensure that the observed resistance in one location is maintained in different geographic regions and resource conditions.

IV. Mite environment

Because varroa is an obligate parasite, the environment of the mite is restricted to the honey bee colony. In the previous section, two examples were presented of how climatic and resource conditions that affect the entire bee colony can also affect the reproductive success of the mites within the colony.

In this section, the discussion will be limited to the more immediate environment of the mite: the adult bees on which the mites are phoretic ("hitchhike"), and the developing larvae and pupae within the cells on which the mites reproduce. Genetic and environmental factors that influence the physiology of the individual adult and/or immature bee may also affect the physiology, and therefore the fertility of the mite.

Several researchers have asked whether increasing the phoretic period of the mites on adult bees decreases the mite's reproductive success (reviewed in Büchler 1994). For example, in winter, or when no brood is available in the nest, the mites are restricted to adult bees. Mites collected from adult bees during the winter were less fertile when introduced into brood cells containing fifth instar larvae than mites transferred during summer months (Hänel and Koeniger 1986). When mites were forced to remain phoretic for six weeks during the summer and there were no young nurse bees in the colony, the reproductive success of the mites also was diminished (Rosenkranz and Bartlaszky 1996). However, when mites were forced to be phoretic for a shorter period of time in summer, 11-20 days, no observed reduction of fertility was observed (Boot et al. 1995).

Other researchers have asked what inhibits the initiation of oviposition once a mite enters a cell containing a fifth instar larva. A fertile female mite will initiate egg-laying approximately 60 hours after the cell is capped. In cases where the mites do not initiate oviposition, it is unclear whether it is due to the status of the mite or to some factor within the brood cell. Hänel and Koeniger (1986) hypothesized that mites do not initiate oviposition on *A. cerana* worker brood because the "trigger" signal from worker brood has a lower stimulus intensity than does the signal from drone brood. The nature of the "trigger" signal remains unknown. It was first postulated that the induction of oogenesis (egg development) by the mite within brood cells was regulated by juvenile hormone (JH III) from the bee larva (Hänel 1983; Hänel and Koeniger 1986). When JH was applied to worker larvae, the reproductive rate of mites collected from bees during the winter increased (Hänel and Koeniger 1986). However, in subsequent studies, JH titers were radioimmunoassayed from fifth instar larvae of *A. m. carnica*, *A. m. lamarckii*, and Africanized bees in temperate and tropical climates (Rosenkranz et al. 1990), and from

fifth instar larvae of *A. cerana* and *A. mellifera* in the same location (Rosenkranz et al. 1993b). In both studies no differences in hormone titres were found among the different bees and it was concluded that JH III could not be the trigger for mite oogenesis.

In a separate study in which comb sections containing brood from six different bee races and four hybrids were introduced into one infested colony, Fuchs (1994) found that the genetic source of the larva did not influence the initiation of oviposition. Additionally, the number of mites that infested an individual cell did not limit the initiation of oviposition, as each female produced a male offspring (the first egg laid in the cell by each female); however, the successful development of the subsequent female offspring was reduced in cells infested by multiple mites (Fuchs and Lagenbach 1989; Martin 1995). Fuchs (1994) concluded the initiation of oviposition is mostly influenced by "unknown colony factors influencing the reproductive state of varroa when they enter cells for reproduction." In a recent study, Nazzi and Milani (1996) speculated that there may be an inhibitor of reproduction released into infested cells such that subsequent invasion of varroa reduces its fertility. In this study, infested pupae were introduced into artificial cell cups in the laboratory; the presence of such an inhibitor has not been confirmed in natural conditions.

The nutritional or hormonal state of the pupae during the *development* of females mites within a cell may influence their fertility as adults. However, the nutritional state of the pupae may be influenced by the nutritional state of the colony (bee environment), genotype of larva (bee genetics), and/or genotype of the mite (mite genetics). Continued research is critical to determine what factors regulate the initiation and successful reproduction of varroa on worker and drone brood. It is necessary to determine if the high proportion of infertile mites observed in some locations is stable over time, or if it is environmentally labile.

V. Current breeding programs

There are a variety of beekeepers and researchers in Europe and the U.S. who are attempting to breed bee colonies that demonstrate resistance to varroa. Some have selected for colonies that display a specific trait such as grooming behavior (Ruttner and Hänel 1992), hygienic behavior (Spivak 1996), and post-capping duration (N. Koeniger, personal communication). Others have bred from

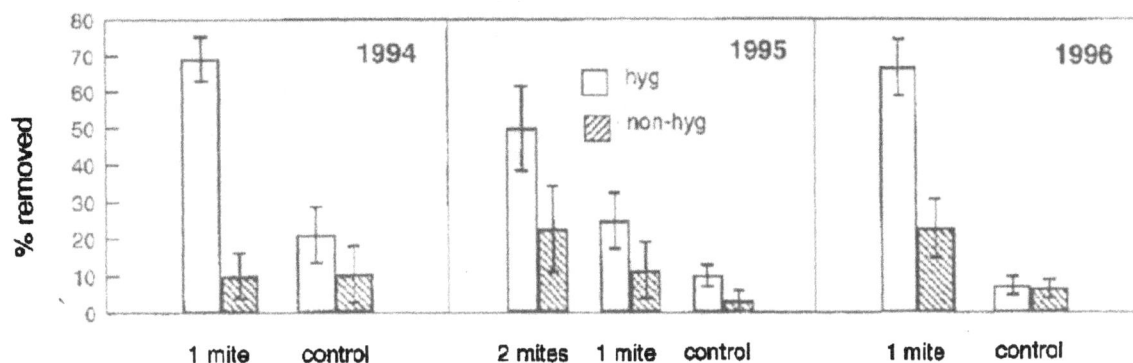

Figure 3. The percent removal of mite-infested pupae by hygienic and non-hygienic colonies 10 days after introduction of the mites through cell bases of Jenter Boxes. In 1994 and 1996, the hygienic colonies (n=4 and 10, respectively) removed significantly more pupae infested with one mite per cell than the non-hygienic colonies (n=3 and 6) (P <0.01; split-plot 2-way ANOVA for each year). There also was a significant difference between the removal of infested pupae and controls (P < 0.05) in both years. Tests in 1995 (n=7 hyg, 4 non-hyg) revealed a significant difference only when two mites per cell were introduced (treatment effect: P < 0.01). Results from 1994 and 1995 in Spivak (1996), and from 1996, unpublished.

colonies that have survived longer with higher infestation levels than other colonies in the same apiaries, *e.g.*, the "Yugo" strain of *A. m. carnica* imported into the U.S. from Yugoslavia (Rinderer *et al.* 1993; de Guzman *et al.* 1996b). Groups of beekeepers have formed cooperatives to breed from colonies that survive without treatment, *e.g.*, "Honey Bee Improvement Program" (J. Griffes, personal communication).

In recent years, some researchers have developed comprehensive approaches to breeding for resistance. In Kirchhain, Germany 200 colonies have been evaluated for mite resistance and performance since 1990 (Büchler 1997). Each colony is infested with 70 varroa in December and a final mite count is taken after a treatment in July to calculate the mite population increase and the proportion of brood and adult bees that are infested. The degree of grooming and hygienic behaviors are evaluated by repeated tests during the season. For selection, the colonies are scored such that the degree of varroa resistance accounts for 50% of the total selection index. Honey yield accounts for 25% of the total, gentleness for 12.5% and calmness on the comb surface, 12.5%. Different breeding lines have demonstrated significantly different degrees of susceptibility and performance using this approach.

Harbo (1996) developed a 10 week procedure for evaluating the population growth of mites and numerically describing which part of the mite's re-

productive cycle is affected by the bee. The procedure involves introducing queens from various sources into colonies that originated from a common source of bees of known infestation level. During the 10 weeks, the following parameters are measured: initial and final mite populations (mite population increase); proportion of brood that is infested; ratio of mites on adults to mites within brood cell; degree of non- reproduction of mites in worker brood cells; post-capping duration of the brood; and degree of grooming and hygienic behaviors. Tests on queens introduced into colonies in Michigan and Baton Rouge in 1996 (Harbo and Hoopingarner 1997) revealed that three of 43 test colonies had fewer mites at the end of the experiment than they had at the beginning. Non-reproduction of mites was the most significant factor related to the observed change in mite population, but had little effect when it occurred at levels <30%. The authors discussed the fact that grooming behavior, hygienic behavior, and post-capping duration may be important mechanisms (see also Danka *et al.* 1997) but "perhaps... are effective only when a colony possesses these traits at a very high level."

To understand the factors responsible for honey bee resistance to varroa infestation, continued research and breeding programs along the lines of Büchler (1997), Harbo (1996), and Harbo and Hoopingarner (1997) are critical. If we breed only from the survivors of untreated colonies, we may obtain colonies with some degree of resistance, but it is important to understand the reasons why some colonies survive. The most efficient breeding program should apply selection pressure on the characteristics that have the greatest impact on reducing mite survival and reproductive success, and those characteristics should be heritable.

Despite some claims to the contrary, there are no beekeepers or researchers who have successfully bred a line of bees that is varroa resistant. Even within the breeding programs for stocks resistant to varroa in North America and Europe, treatments are absolutely necessary at present to keep the colonies alive. In the search for bee colonies that are resistant to varroa, beekeepers should consider the following question: How long should we expect colonies to survive without chemical treatment? Should we be selecting for colonies that never require the application of miticides, or that survive without treatment for one or two years? Breeding for bee colonies that survive varroa infestation without any chemical control is the most sustainable solution to the problem, but to achieve that end, we must continue to search for the most critical factors that limit the survivorship and reproductive success of the varroa within honey bee colonies.

Acknowledgments

We thank Jeremiah Bowman, John Harbo, and Rebecca Melton for reviewing this manuscript.

References

Anderson, D.L. 1994. Non-reproduction of *Varroa jacobsoni* in *Apis mellifera* colonies in Papua New Guinea and Indonesia. *Apidologie* 25: 412-421.

Anderson, D.L. and Sukarsih. 1996. Changed *Varroa jacobsoni* reproduction in *Apis mellifera* colonies in Java. *Apidologie* 27: 461-466.

Anderson, D.L. and S. Fuchs. 1998. Genetically distinct populations of *Varroa jacobsoni* with contrasting reproductive abilities on the European honey bee, *Apis mellifera J. Apic. Res.* In press.

Boecking, O. 1992. Removal behaviour of *Apis mellifera* colonies towards sealed brood cells infested with *Varroa jacobsoni*: techniques, extent and efficacy. *Apidologie* 23: 371-373.

Boecking, O. 1994. The removal behavior of *Apis mellifera* L. towards mite-infested brood cells as an defense mechanism against the ectoparasitic mite *Varroa jacobsoni* Oud.. *Ph.D. thesis, Rheinische-Friedrich-Wilhelms-Universität*, Bonn. 127 p.

Boecking, O. and W. Drescher. 1990. The reaction of worker bees in different *Apis mellifera* colonies to Varroa infested brood cells. pp. 41-42 *in* Ritter, W. (ed)

Proceedings of the international symposium on recent research on bee pathology, Sept. 1990, Gent, Belgium; 223 p.

Boecking, O. and W. Drescher. 1991. Response of *Apis mellifera* L. colonies to brood infested with *Varroa jacobsoni.* Oud. *Apidologie* 22: 237-241.

Boecking, O. and W. Drescher. 1992. The removal response of *Apis mellifera* L. colonies to brood in wax and plastic cells after artificial and natural infestation with *Varroa jacobsoni* Oud. and to freeze-killed brood. *Exp. Appl. Acarol.* 16: 321-329.

Boecking, O. and W. Drescher. 1993. Preliminary data on the response of *Apis mellifera* to brood infested with *Varroa jacobsoni* and the effect of this resistance mechanism. pp. 454-462. *in* Connor, L.J., T. Rinderer, H.A. Sylvester, and S. Wongsiri (eds) *Asian Apiculture.* Wicwas Press, Cheshire, USA. 704 p.

Boecking, O. and W. Ritter. 1993. Grooming and removal behaviour of *Apis mellifera intermissa* in Tunisia against *Varroa jacobsoni. J. Apic. Res.* 32: 127-134.

Boecking, O. and W. Ritter. 1994. Current status of behavioral tolerance of the honey bee *Apis mellifera* to the mite *Varroa jacobsoni. Am. Bee J.* 134: 689 -694.

Boecking, O., and M. Spivak. Behavioral defenses of honey bees against *Varroa jacobsoni* Oud *Apidologie.* In press.

Boot, W.J., J. N. M. Calis and J. Beetsma. 1995. Does time spent on adult bees affect reproductive success of *Varroa* mites. *Entomol. Exp. et Appl* 75: 1-7.

Büchler, R. 1994. Varroa tolerance in honey bees - occurrence, characters and breeding. *Bee World* 75: 54-70.

Büchler, R. 1997. Aktuelle ergebnisse zur selektion auf Varroatoleranz. *Allg. Deut. Emkerzeitung* 31: 10-15.

Büchler, R. and W. Drescher. 1990. Variance and heritability of the capped developmental stage in European *Apis mellifera* L. and its correlation with increased *Varroa jacobsoni* Oud. infestation. *J. Apic. Res.* 29: 172-176.

Büchler, R., W. Drescher and I. Tournier. 1992. Grooming behaviour of *Apis cerana, Apis mellifera* and *Apis dorsata*, reacting to *Varroa jacobsoni* an *Tropilaelaps clareae. Exp. Appl. Acarol.* 16: 313-319.

Camazine, S. 1986. Differential reproduction of the mite, *Varroa jacobsoni* (Mesostigmata: Varroidae), on Africanized and European honey bees (Hymenoptera: Apidae). *Annals Entomol Soc Amer.* 79: 801-803.

Danka, R.G., J.D. Villa, J.R. Harbo and T.E. Rinderer. 1997. Initial evaluation of industry-contributed honey bees for resistance to *Varroa jacobsoni. Amer. Bee J.* Proceedings of ABRC. 137: 221-222.

de Guzman, L.I., T.E. Rinderer and J.A. Stelzer. 1996b. Determination of the origin of *Varroa jacobsoni* in the Americas using RAPD. Abstract for 3rd Asian Apicultural Assoc. Conf. Oct 6-10, 1996, Hanoi, Vietnam.

de Guzman, L.I., T.E. Rinderer, G.T. Delatte and R.E. Macchiavelli. 1996b. *Varroa jacobsoni* Oudemans tolerance in selected stocks of *Apis mellifera* L. *Apidologie* 27: 193-210.

DeJong, D., L.S. Gonçalves and R.A. Morse. 1984. Dependence on climate of the virulence of *Varroa jacobsoni. Bee World* 65: 117-121.

Engels, W., L.S. Gonçalves, J. Steiner, A.M. Buriolla and M.R. Cavichio Issa. 1986. *Varroa-Befall von Carnica-*

Völkern in Tropenklima. *Apidologie* 17: 203-216.

Fries, I., S. Camazine and J. Sneyd. 1994. Population dynamics of *Varroa jacobsoni*: a model and a review. *Bee World* 75: 5-28.

Fries, I., H. Wei, W. Shi and S.X. Huazhen. 1996. Grooming behavior and damaged mites (*Varroa jacobsoni*) in *Apis cerana cerana* and *Apis mellifera ligustica*. *Apidologie* 27: 3-11.

Fuchs, S. 1992. Choice in *Varroa jacobsoni* Oud. between honey bee drone or workerbrood cells for reproduction. *Behav. Ecol. Sociobiol.* 31: 429-435.

Fuchs, S. 1994. Non-reproducing *Varroa jacobsoni* Oud. in honey bee worker cells - status of mites or effect of brood cells? *Exp. & Appl. Acarology* 18: 309-317.

Fuchs, S. and K. Lagenbach. 1989. Multiple infestation of *Apis mellifera* L. brood cells and reproduction in *Varroa jacobsoni* Oud. *Apidologie* 20: 257-266.

Gilliam, M., S. Taber III and G.V. Richardson. 1983. Hygienic behavior of honey bees in relation to chalkbrood disease. *Apidologie* 14: 29-39.

Gilliam, M., S. Taber III, B. J. Lorenz and D. B. Prest. 1988. Factors affecting development of chalkbrood disease in colonies of honey bees, *Apis mellifera*, fed pollen contaminated with *Ascosphaera apis*. *J. Invertebr. Pathol.* 52: 314-325.

Guzman-Novoa, E., A. Sanchez, R.E. Page, Jr. and T. Garcia 1996 Susceptibility of European and Africanized honeybees (*Apis mellifera* L.) and their hybrids to *Varroa jacobsoni* Oud. *Apidologie* 27: 93-103.

Hänel, H. 1983. Effect of JH-III on the reproduction of *Varroa jacobsoni*. *Apidologie* 14: 137-142.

Hänel, H. and N. Koeniger. 1986. Possible regulation of the reproduction of the honey bee mite *Varroa jacobsoni* (Mesostigmata: Acari) by a hosts hormone: Juvenile hormone III. *J. Insect Physiol.* 32: 791-798.

Harbo, J.R. 1996. Evaluating colonies of honey bees for resistance to *Varroa*. *Bee Sci.* In press.

Harbo, J.R. and R.A. Hoopingarner. 1997. Honey bees (Hymenoptera: Apidae) in the United States that express resistance to *Varroa jacobsoni* (Mesostigmata: Varroidae). *J. Econ. Entomol.* 90:893-898.

Koeniger, N., G. Koeniger and N.H.P. Wijayagunasekara. 1981. Beobachtungen über die Anpassung von *Varroa jacobsoni* an ihren natürlichen Wirt *Apis cerana* in Sri Lanka. *Apidologie* 12: 37-40.

Koeniger, N., G. Koeniger and M. Delfinado-Baker. 1983. Observations on mites of the Asian honeybee species. *Apidologie* 14:197-204.

Kulinçeviç, J.M., T.E. Rinderer, D.J. Uroseviç. 1988. Seasonality and colony variation of reproducing and non-reproducing *Varroa jacobsoni* females in western honey bees (*Apis mellifera*) worker brood. *Apidologie* 19: 173-179.

Le Conte, Y., J.M. Cornuet. 1989. Variability of the postcapping stage duration of the worker brood in three different races of *Apis mellifera*. *in* Cavalloro, R. (ed) *Present status of varroatosis in Europe and progress in the Varroa mite control*. Commission of the European communities; Luxembourg; pp. 171-175.

Martin, S. J. 1995. Reproduction of *Varroa jacobsoni* in cells of *Apis mellifera* containing one or more mother mites and the distribution of these cells. *J. Apic. Res.* 34: 187-196.

Monaco, R. 1997. Development of resistance against Varroa jacobsoni Oudemans in a natural population of Apis mellifera ligustica Spinola. *Am. Bee J.* 137: 140-1142.

Moosbeckhofer, R. 1992. Beobachtungen zum Auftreten beschädigter Varroamilben im natürlichen Totenfall von *Apis mellifera carnica*. *Apidologie* 23: 523-531.

Moretto, G., L.S. Gonçalves and D. De Jong. 1991a. Africanized bees are more efficient at removing *Varroa jacobsoni* - preliminary data. *Am. Bee J.* 131: 434.

Moretto, G., L.S. Gonçalves, D. De Jong and M.Z. Bichuette. 1991b. The effects of climate and bee race on *Varroa jacobsoni* Oud. infestations in Brazil. *Apidologie* 22: 197-203.

Moretto, G., L.S. Gonçalves and D. De Jong. 1993. Heritability of Africanized and European honey bee defensive behavior against the mite *Varroa jacobsoni*. *Rev. Brasil. Genet.* 16: 71-77.

Moretto, G., L.S. Goncalves and D. De Jong. 1997. Relationship between food availability and the reproductive ability of the mite *Varroa jacobsoni* in Africanized bee colonies. *Am. Bee J.* 137: 67-69.

Moritz, R.F.A. 1985. Heritability of the postcapping stage in *Apis mellifera* and its relation to varroatosis resistance. *J. Heredity* 76: 267-270.

Moritz, R.F.A. and H. Hänel. 1984. Restricted development of the parasitic mite *Varroa jacobsoni* Oud. in the Cape honey bee *Apis mellifera capensis* Esch.*Z. angew. Ent.* 97: 91-95.

Moritz, R.F.A. and D. Mautz. 1990. Development of *Varroa jacobsoni* in colonies of *Apis mellifera capensis* and *Apis mellifera carnica*. *Apidologie* 21: 53-58.

Nazzi, F. and N. Milani. 1996. The presence of inhibitors of the reproduction of *Varroa jacobsoni* Oud. (Gamasida: Varroidae) in infested cells. *Exp. & Appl. Acarology* 20: 617-623.

Otten, C. 1991. Factors and effects of a different distribution of *Varroa jacobsoni* between adult bees and bee brood. *Apidologie* 22: 465-467.

Otten, C. and S. Fuchs. 1990. Seasonal variations in the reproductive behavior of *Varroa jacobsoni* in colonies of *Apis mellifera carnica*, *A. m. ligustica* and *A. m. mellifera*. *Apidologie* 21: 367-368.

Peng, Y.S., Y. Fang, S. Xu and L. Ge. 1987a. The resistance mechanism of the Asian honey bee, *Apis cerana* Fabr., to an ectoparasitic mite *Varroa jacobsoni* Oudemans. *J. Invertebr. Pathol.* 49: 54-60.

Peng, Y.S., Y. Fang, S. Xu, L. Ge and M.E. Nasr. 1987b. Response of foster Asian Honey bee (*Apis cerana* Fabr.) colonies to the brood of European honey bee (*Apis mellifera* L.) infested with parasitic mite *Varroa jacobsoni* Oudemans. *J. Invertebr. Pathol.* 49: 259-264.

Rath, W. 1991. Investigations on the parasitic mites *Varroa jacobsoni* Oud. and *Tropilaelaps clareae* Delfinado & Baker and their hosts *Apis cerana* Fabr., *Apis dorsata* Fabr. and *Apis mellifera* L.. Ph.D. thesis, Rheinische-Friedrich-Wilhelms-Universität, Bonn. 148 p.

Rath, W. 1992. The key to Varroa: The drones of *Apis cerana* and their cell cap. *Am. Bee J.* 132: 329-

331.

Rath, W. 1993. Aspects of preadaptation in *Varroa jacobsoni* while shifting from its original host *Apis cerana* to *Apis mellifera*. pp. 417-426. *in* Connor, L.J., T. Rinderer, H.A. Sylvester, and S. Wongsiri (eds) *Asian Apiculture*. Wicwas Press, Cheshire, USA. 704 p.

Rath, W. and W. Drescher. 1990. Response of *Apis cerana* Fabr. towards brood infested with *Varroa jacobsoni* Oud. and infestation rate of colonies in Thailand. *Apidologie* 21: 311-321.

Rinderer, T.E., L.I. de Guzman, J.M. Kulinçeviç, G.T. Delatte, L.D. Beaman, and S.M. Buco. 1993. The breeding, importing, testing and general characteristics of Yugoslavian honey bee bred for resistance to *Varroa jacobsoni. Am. Bee J.* 133: 197-200.

Ritter, W. 1990. Development of the Varroa mite populations in treated and untreated colonies in Tunisia. *Apidologie* 21: 368-370.

Ritter, W. 1993. New results of the development of tolerance to *Varroa jacobsoni* in bee colonies in Tunisia. pp. 463-467. *in* Connor, L.J., T. Rinderer, H.A. Sylvester, and S. Wongsiri (eds) *Asian Apiculture*. Wicwas Press, Cheshire, USA. 704 p.

Ritter, W., P. Michel, A. Bartholdi and A. Schwendemann. 1990. Development of tolerance to *Varroa jacobsoni* in bee colonies in Tunisia. pp. 54-59. *in* Ritter, W., O. van Laere, F. Jacobs, and L. de Wael (eds) *Proc. Recent Research on Bee Pathology.* Apimondia 1990, Gent, Belgium. 223 p.

Rosenkranz, P. and H. Bartlaszky. 1996. Reproduction of *Varroa* females after long broodless periods of the honey bee colony during summer. German Bee Research Institues Seminar. *Apidologie* 27: 288-289, (abstract)

Rosenkranz, P. and W. Engels. 1994. Infertility of *Varroa jacobsoni* females after invasion into *Apis mellifera* worker brood as a tolerance factor against varroatosis. *Apidologie* 25: 402-411.

Rosenkranz, P., A. Rachinsky, A. Strambi, C. Strambi and P. Röpstorf. 1990. Juvenile hormone titer in capped worker brood of *Apis mellifera* and reproduction in the bee mite *Varroa jacobsoni. Gen. Comp. Endocrinol.* 78: 189-193.

Rosenkranz, P. and N.C. Tewarson. 1992. Experimental infection of *Apis cerana indica* worker brood with Varroa females. *Apidologie* 23: 365-367.

Rosenkranz, P., N.C. Tewarson, A. Singh and W. Engels. 1993a. Differential hygienic behaviour towards *Varroa jacobsoni* in capped worker brood of *Apis cerana* depends on alien scent adhering to the mites. *J. Apic. Res.* 32: 89-93.

Rosenkranz, P., N.C. Tewarson, A. Rachinsky, A. Strambi, C. Strambi and W. Engels. 1993b. Juvenile hormone titer and reproduction of *Varroa jacobsoni* in capped brood stages of *Apis cerana indica* in comparison to *Apis mellifera ligustica. Apidologie* 24: 375-382.

Rothenbuhler, W. 1964. Behavior genetics of nest cleaning behavior in honeybees. I. Response of four inbred lines to disease killed brood. *Animal Behavior* 12: 578-583.

Ruttner, F., H. Marx and G. Marx. 1984. Beobachtungen über eine mögliche Anpassung von *Varroa jacobsoni* an *Apis mellifera* L. in Uruguay. *Apidologie* 15: 43-62.

Ruttner, F. and H. Hänel. 1992. Active defense against Varroa mites in Carniolan strains of honey bees. *Apidologie* 23: 173-187.

Schousboe, C. 1986. The duration of sealed cell stage in worker honeybee brood (*Apis mellifera* L.) in relation to increased resistance to the Varroa mite (*Varroa jacobsoni* Oud.). *Tidsskrift for Planteavl* 90: 293-299.

Spivak, M. and M. Gilliam. 1993. Facultative expression of hygienic behaviour of honey bees in relation to disease resistance. *J. Apic. Res.* 32: 147-157.

Spivak, M. 1996. Hygienic behavior and defense against *Varroa jacobsoni.* Apidologie 27: 245-260.

Spivak, M. and D. Downey. 1998. Field assays for hygienic behavior in honey bees (Apidae: Hymenoptera). *J. Econ. Entomol.* 91: 64-70.

Spivak, M. and G.A. Reuter. 1998. Performance of Hygienic Colonies in a Commercial Apiary. *Apidologie.* 29: 285-290.

Spivak, M. and O. Boecking. Honey bee resistance to *Varroa jacobsoni* mites. *In* T. Webster and K. Delaplane, (Ed.). In press.

Taber, S. and M. Gilliam. 1987. Breeding honey bees for resistance to diseases. *Korean J. Apic.* 2: 15-20.

Takeuchi, K. 1993. Extinction of *Varroa* mites in Japanese honeybee (*Apis cerana japonica*) colony. *Honeybee Science* 14: 58-60.

Tewarson, N.C. 1987. Use of host hemolymph proteins, seasonal reproduction and a hypothesis on nutritional imprinting in the honey bee mite, *Varroa jacobsoni,* on *Apis mellifera* and *Apis cerana. in* Eder, J. and H. Rembold (eds) *Chemistry and Biology of Social Insects.* J. Peperny, Munich, 688-689.

Tewarson, N.C., A. Singh and W. Engels. 1992. Reproduction of *Varroa jacobsoni* in colonies of *Apis cerana indica* under natural and experimental conditions. *Apidologie* 23: 161-171.

Wallner, A. 1990. Beobachtungen natürlicher Varroa-Abwehrreaktionen in meinen Bienenvölkern. *Imkerfreund* 9: 4-5.

Woodrow, A.W. and E. C. Holst. 1942. The mechanism of colony resistance to American foulbrood. *J. Econ. Entomol.* 35: 327-330.

Key words: *Varroa jacobsoni, Apis mellifera,* reproductive success, hygienic behavior

Index